Angelina Birchler Pedross

Circadian and sleep-wake homeostatic processes in depressed women

Angelina Birchler Pedross

Circadian and sleep-wake homeostatic processes in depressed women

Influences on mood and higher frontal EEG synchronization during sustained wakefulness

Südwestdeutscher Verlag für Hochschulschriften

Impressum / Imprint
Bibliografische Information der Deutschen Nationalbibliothek: Die Deutsche Nationalbibliothek verzeichnet diese Publikation in der Deutschen Nationalbibliografie; detaillierte bibliografische Daten sind im Internet über http://dnb.d-nb.de abrufbar.
Alle in diesem Buch genannten Marken und Produktnamen unterliegen warenzeichen-, marken- oder patentrechtlichem Schutz bzw. sind Warenzeichen oder eingetragene Warenzeichen der jeweiligen Inhaber. Die Wiedergabe von Marken, Produktnamen, Gebrauchsnamen, Handelsnamen, Warenbezeichnungen u.s.w. in diesem Werk berechtigt auch ohne besondere Kennzeichnung nicht zu der Annahme, dass solche Namen im Sinne der Warenzeichen- und Markenschutzgesetzgebung als frei zu betrachten wären und daher von jedermann benutzt werden dürften.

Bibliographic information published by the Deutsche Nationalbibliothek: The Deutsche Nationalbibliothek lists this publication in the Deutsche Nationalbibliografie; detailed bibliographic data are available in the Internet at http://dnb.d-nb.de.
Any brand names and product names mentioned in this book are subject to trademark, brand or patent protection and are trademarks or registered trademarks of their respective holders. The use of brand names, product names, common names, trade names, product descriptions etc. even without a particular marking in this works is in no way to be construed to mean that such names may be regarded as unrestricted in respect of trademark and brand protection legislation and could thus be used by anyone.

Coverbild / Cover image: www.ingimage.com

Verlag / Publisher:
Südwestdeutscher Verlag für Hochschulschriften
ist ein Imprint der / is a trademark of
AV Akademikerverlag GmbH & Co. KG
Heinrich-Böcking-Str. 6-8, 66121 Saarbrücken, Deutschland / Germany
Email: info@svh-verlag.de

Herstellung: siehe letzte Seite /
Printed at: see last page
ISBN: 978-3-8381-3354-6

Zugl. / Approved by: Zürich, UZH, Diss., 2012

Copyright © 2012 AV Akademikerverlag GmbH & Co. KG
Alle Rechte vorbehalten. / All rights reserved. Saarbrücken 2012

TABLE OF CONTENTS

ZUSAMMENFASSUNG/SUMMARY   3

CHAPTER 1
Theoretical Background   7
Objectives and Methods   17
Outline of the thesis   19
Aims and hypothesis   20

CHAPTER 2
Subjective Well-being is modulated by Circadian Phase,
Sleep Pressure, Age, and Gender   22

CHAPTER 3
Higher Frontal EEG Synchronization in Young Women
With Major Depression: A Marker for increased Homeostatic
Sleep Pressure?   41

CHAPTER 4
Sleep Deprivation and Diurnal Variations of Mood in
Untreated Unipolar Depressed Women under High and
Low Sleep Pressure Conditions   57

CHAPTER 5
Young Depressed Women perform faster in a
Psychomotor Vigilance Task during Sleep Deprivation
than Controls   68

CHAPTER 6
General Discussion   80

REFERENCES   87
List of Figures and List of Tables   107

ZUSAMMENFASSUNG

In dieser Doktorarbeit wird die Depression im Rahmen der zirkadianen und homöostatischen Schlaf-Wach-Regulation untersucht. Wie wirken sich diese zwei unabhängigen und nicht additiven Systeme auf die Stimmung, die subjektive Schläfrigkeit, subjektive Anspannung, das Wach-EEG, die Melatoninkonzentration, und die kognitive Leistungsfähigkeit bei jungen depressiven Frauen ohne Insomnie unter zwei verschiedenen Schlafdruckbedingungen aus? Verschiedene Untersuchungen weisen auf eine Beteiligung respektive einer ungünstigen Interaktion der Schlafhomöostase und des endogenen zirkadianen Systems bei der Entstehung von depressiven Erkrankungen hin. Überraschenderweise wurden bis jetzt sehr wenige Studien unter standardisierten kontrollierten Laborbedingungen bei der Majoren Depression (MDD) durchgeführt.

Das Studiendesign besteht aus zwei verschiedenen Protokollen und wurde unter „Constant Routine" - Bedingungen durchgeführt; beide beginnen mit einer Baselinenacht und schliessen mit einer Erholungsnacht ab. Die 40 Stunden zwischen diesen beiden Nächten bestehen entweder aus einer Episode mit totaler Schlafdeprivation (SD; hoher Schlafdruck) oder aus 10 Schlaf/Wach-Zyklen von 75 min. Schlaf, gefolgt von 150 min. Wachsein (Nap Protokoll; tiefer Schlafdruck).

Das erste Ziel dieser Doktorarbeit (Kapitel 2) war die Untersuchung der zirkadianen und schlaf-wach-homöostatischen Veränderungen in der subjektiven Befindlichkeit an gesunden Menschen. Die Interaktion dieser beiden Prozesse wurde unter zwei verschiedenen Schlafdruckbedingungen mit der Frage nach Geschlechts- und Altersunterschieden untersucht. Generell zeigten Frauen eine schlechtere subjektive Befindlichkeit und ein markanteres zirkadianes Stimmungstief unter Schlafdeprivation. Im Vergleich zu den Jüngeren wurde bei älteren Menschen unter diesen Studienbedingungen eine schlechtere subjektive Befindlichkeit gemessen. Diese Resultate zeigen, dass sich das zirkadiane System sowie die Schlaf-Wach-Homöostase geschlechts- und alterspezifisch auf die subjektive Befindlichkeit auswirken.

In einem weiteren Schritt wurde die Auswirkung der Interaktion der zirkadianen und schlaf-wach-homöostatischen Prozessen auf verschiedene Parameter bei jungen Frauen mit einer Episode einer Majoren Depression unter hohen und tiefen Schlafdruckbedingen untersucht.

Die Hypothese, dass Depression auf einen Mangel im schlaf-wach-homöostatischen Prozess zurückführt (d.h. die S-Defizient-Hypothese) wurde im zweiten Beitrag (Kapitel 3) mittels Wach-EEG während anhaltender Wachheit überprüft. Gleichzeitig wurde die subjektive Schläfrigkeit, subjektive Anspannung und die Melatoninkonzentration im Speichel untersucht. Im Gegensatz zur postulierten Defizienz des homöostatischen Prozesses („Schlafdruckmanko"), zeigten sich erhöhte langsamwellige EEG Anteile (Deltaband von 2 bis 5 Hz) während des Wachseins (SD Protokoll) vor allem in frontalen Hirnregionen mit gleichzeitig erhöhter Schläfrigkeit und subjektiver Anspannung sowie einer verringerten nächtlichen Melatoninausschüttung. Diese Resultate lassen vermuten, dass depressive Frauen auf einem höheren Schlafdruckniveau „leben". Somit könnte eher von einer Übersteuerung der Schlaf-Wach-Homöostase gesprochen werden.

Im dritten Beitrag (Kapitel 4) wird der Fokus auf tageszeitliche Stimmungsschwankungen in der Depression gelegt. Ausgehend von den Kenntnissen über Schlafentzugsbehandlungen bei Depressiven wurde bei den jungen depressiven Frauen ein möglicher stimmungsaufhellender Effekt unter hohen Schlafdruckbedingungen untersucht. Die Daten zeigen, dass depressive Frauen ohne Schlafstörungen im Schlafdeprivationsprotokoll einen signifikant unterschiedlicheren Stimmungsverlauf im Vergleich zu den gesunden jungen Frauen zeigen. Trotz dieser ausgeprägten Stimmungsschwankung konnten die jungen depressiven Frauen nicht von einem antidepressiven Effekt profitieren. Diese Ergebnisse weisen darauf hin, dass der mögliche Profit einer Schlafdeprivationsbehandlung vom Ausmass des Vorhandenseins der Schlafstörung bei Depressiven abhängig ist.

Im letzten Beitrag (Kapitel 5) wurde der Einfluss von hohem und tiefem Schlafdruck auf die kognitive Leistung bei Depressiven mittels Psychomotorischen Vigilanztest (PVT) untersucht. Unerwarteterweise zeigte sich bei den jungen depressiven Frauen eine schnellere Reaktionszeit im PVT während der Schlafdeprivation. Dies könnte auf eine geringere Anfälligkeit auf wachheitsabhängige nachteilige Aspekte bei der Leistung des PVT schliessen lassen.

Aus den Ergebnissen dieser Arbeit lässt sich schliessen, dass in der Interaktion des zirkadianen und schlaf-wach-homöostatischen Prozesses unter streng kontrollierten Bedingungen und bei einem mittleren Chronotyp, depressive junge Frauen deutliche Unterschiede zu den gesunden jungen Frauen aufweisen. Die Resultate im schlaf-wach-homöostatischen Prozess sind anders als erwartet, und haben in der Interaktion möglicherweise einen grösseren Einfluss als bisher angenommen.

Für die Behandlung der Depression könnte aufgrund dieser Resultate die Verabreichung von Licht am Morgen zur Reduktion der Schläfrigkeit und der Erhöhung der Melatoninamplitude hilfreich sein.

SUMMARY

There is mounting evidence supporting the role of the sleep-wake cycle and the endogenous circadian system in the pathogenesis of the disorder of major depression. Surprisingly, very little research has been done so far with the focus on basic investigations under unmasked standard controlled conditions to avoid masking effects in Major Depression Disorder (MDD). In this thesis, the theoretical framework that underpins depression is addressed within the context of the circadian and sleep-wake homeostatic regulation, and on how these two independent and non-additive systems impact mood, subjective sleepiness, melatonin and waking EEG in young depressed women without sleep disturbances.

The study design consisted of two different protocols; both started with a baseline and ended with a recovery night. The 40-h episode between these two nights comprised either an episode of total sleep deprivation (SD; high sleep pressure) or 10 sleep/wake cycles with 75 min of sleep followed by 150 min of wakefulness (nap protocol; low sleep pressure). In this thesis the focus is on the response of the different parameters of the episodes during the SD protocol and, furthermore, a comparison is made of the courses of both protocols.

The first aim of this thesis was to gain comprehensive information about a basic model to quantify circadian and sleep-wake related changes in subjective and objective variables in healthy volunteers to understand what normal daily variations of this parameters look like. In this regard, subjective well-being, subjective sleepiness, melatonin and cortisol levels (Chapter 2) were first compared among two different age and gender groups of healthy participants under unmasked conditions in a high and low sleep pressure protocol. The time course of subjective well-being shows a circadian rhythm under low and high sleep pressure conditions. Furthermore, we could show that the well known circadian regulation of subjective sleepiness, melatonin and cortisol are clearly age and gender dependent in healthy adult volunteers. Time of day modulation of subjective well-being was prominent in both the SD and the nap protocol, indicating that circadian phase plays a pivotal role in well-being. In general, and as a new interesting finding, both older adults and women were more affected by SD, showing a tendency to lower subjective well-being and a prominent circadian trough. The time course of subjective well-being displayed a significant circadian modulation, particularly in women under high sleep pressure conditions.

The hypothesis that depression could be linked to a deficiency in the sleep-wake homeostatic process (i.e. the so-called S-deficiency hypothesis) was tested by waking EEG recordings during sustained wakefulness along with assessment of subjective sleepiness, tension and salivary melatonin (Chapter 3). In contrast to a deficiency of the homeostatic process (i.e. S-deficiency), our results indicated that depressive women might live on a higher level of homeostatic sleep pressure, as indexed by enhanced high frontal-low EEG activity (delta range from 2 to 5 Hz) during sustained wakefulness in the SD protocol. This finding of a higher sleep pressure in depressives was also supported with sleep recording results that showed a tendency to have higher sleep pressure levels during the nap protocol. Along these lines, we have evidence for significantly higher EEG slow-wave activity (delta EEG activity) levels in Non-REM (NREM) sleep in our depressed cohort during baseline and recovery sleep in both the SD and the nap protocol.

Chapter 4 focuses on diurnal variations of mood and whether there is an antidepressant effect of SD. The data show that depressive women without sleep disturbances show a significantly different time course of mood during the SD protocol but not during nap protocol. Under SD they exhibited a more distinct circadian modulation of lower mood than controls. In their diurnal modulation they showed morning worsening and evening improvement, which corresponds to the so-called melancholic type. Despite this higher variability of mood fluctuation and characteristic of a more melancholic type as a predictor of SD response, surprisingly and in contrast to our hypothesis, they did not profit from an antidepressant effect. These findings indicate that the possible benefit of SD treatment could depend on the magnitude of insomnia in depression.

In a final step, in Chapter 5, we focused on the influence of high and low sleep pressure conditions on neurobehavioral performance in depressed patients using the psychomotor vigilance task (PVT). These results yielded an unexpected finding: depressed women had faster reaction times (RT) in PVT performance during SD. This may imply that depressed women are less susceptible to the wake-dependent aspects in PVT performance.

Taken together, this thesis provides several surprising data in our cohort of MDD women that stand in contrast to some of the long-standing mainstream hypothesis (chapters 3–5). Overall, the homeostatic process seems to override young women with major depression disorder, as indicated by enhanced high frontal low-EEG activity. The depressed reported higher subjective sleepiness and had lower salivary melatonin, detected during a biological night, indicating that they did not profit from the antidepressant effect yet performed faster in PVT during SD. The waking EEG data suggests that the sleep-wake homeostat of the depressed is more an overdrive than an S-deficiency, which probably corresponds to the clinically reported day time fatigue. Depressed women who live on a higher level of sleep pressure do not profit from the antidepressant effect of SD and perform faster in a reaction time task than controls. This leads to the assumption that the excluded factors in this protocol (chronotype, insomnia) may exert a strong influence on these results. The suggestion that hypocretin neurons stabilise arousal/alertness during periods of wakefulness and increase arousal-related behaviours can be interpreted as having an endogenous stressor effect. The heuristic model of emotional and physiological hyperarousal hypothesis must be assumed, and the depressive cohort in our study suffered only a minor hyperarousal, which probably affects frontal delta activity, leading to a higher sleep pressure but is not strong enough to cause sleep disturbances. The postulated abnormalities (i.e. phase-shift, sleep disturbances) in the biological rhythms seem to undergo less dramatic differences in the young depressed women who were investigated in this thesis; hence, it can be assumed that the homeostatic process seems to have more influence than previously thought.

CHAPTER 1

Theoretical Background

Introduction

There is mounting evidence supporting the role of the sleep-wake cycle and the endogenous circadian system in the pathogenesis of the disorder of major depression. In general, several psychiatric disorders like unipolar depression, bipolar disorder, seasonal affective disorder and anxiety-related disorders have been associated with circadian abnormalities (Boivin 2000; McClung 2007; Roybal, Theobold et al. 2007). Disturbed circadian rhythms such as diurnal variation of mood and sleep belong to the major symptoms of depressive patients. The first studies of clock genes in major depression were negative (Desan, Oren et al. 2000). Thus, it is still not clear if alterations in circadian clock gene expression play a significant role in depression. Disorders of the human circadian system can result in circadian misalignment, which itself causes sleep disturbances, reduced attention, impaired daytime alertness, fatigue, lack of energy, memory problems, abnormal negative mood and gastrointestinal disorders. Interestingly, all these neuropsychiatric symptoms also occur in depression and are classified in the international diagnosis systems.

Surprisingly very little research has been done so far with the focus on basic investigations under unmasked standard controlled conditions to avoid masking effects in depression. Most clinical and basic investigations on circadian alterations in depression have focussed on the subgroup of Seasonal Affective Disorder (SAD). From these studies, concepts of "susceptible circadian phases", during which sudden shifts in circadian rhythms lead to depression in predisposed individuals, was initially proposed by Papousek and colleagues (Papousek 1975) and refined by Wehr and Wirz-Justice (Wehr and Wirz-Justice 1982) and others (see for review (Wirz-Justice 1995)). Earlier sleep laboratory investigations yielded an association between abnormal sleep patterns including shortened REM latency and early morning awakening with major depression (Gillin, Duncan et al. 1979). However, newer sleep studies including the spectral analyses of the sleep EEG in depression could not confirm earlier data (Koorengevel, Beersma et al. 2002) (Landolt and Gillin 2005).

The two-process models of sleep regulation

Sleep homeostasis regulates the balance between sleep and waking. Two interacting processes, a homeostatic process S and a circadian process C have been postulated to regulate sleep and wakefulness (Borbely 1982; Daan, Beersma et al. 1984). Biological rhythms are periodical temporal variations, statistically proved, foreseeable and recognizable. The two main processes characterizing the sleep-wake regulation are; a homeostatic process S describing the sleep-pressure's course during wake-time and its lowering during sleep; a circadian process C with a period length of about 24 hours. Both processes influence a third one, the so-called ultradian process, which regulates the structure of sleep cycles, presented in Figure 1.1.

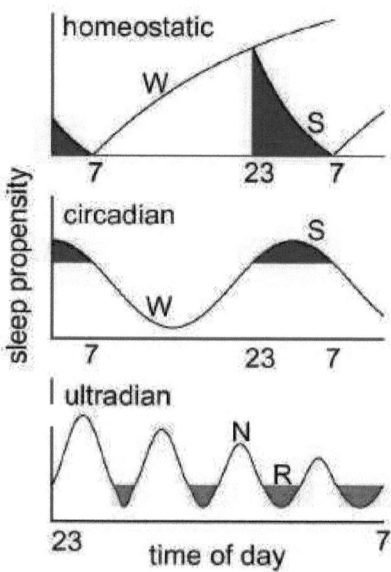

Figure 1.1: Schematic representation of the homeostatic and circadian processes; as well as the ultradian process during sleep: W, waking; S, sleep; N, NREM sleep; R, REM sleep. The progressive decline of NREM intensity is represented both in the top and bottom diagrams (decline of ultradian amplitude). The increase in the duration of successive REM sleep episodes is indicated (Figure modified from (Achermann and Borbely 1992)).

The sleep homeostatic process

Process S, the homeostatic pressure for sleep progressively increases as an exponential saturating function during waking and dissipates as an exponential function during the following sleep episode. The extent of homeostatic sleep pressure directly depends on the duration of prior wakefulness (Borbely, Baumann et al. 1981; Dijk, Beersma et al. 1987). Therefore, S represents an hourglass process. The term "sleep homeostasis" (Borbely, Steigrad et al. 1980) was proposed to characterize the sleep-wake dependent aspect of sleep regulation. Characteristics of the homeostatic process as a measurement are low frequency components such as EEG theta activity (4.5-8 Hz) during wakefulness (Cajochen, Khalsa et al. 1999; Finelli, Baumann et al. 2000; Cajochen,

Knoblauch et al. 2001) and the EEG slow-wave activity (SWA; 0.75-4.5 Hz) during sleep (Borbely, Baumann et al. 1981), (Werth, Achermann et al. 1997; Cajochen, Foy et al. 1999; Finelli, Borbely et al. 2001). Changes in homeostatic sleep regulation can be quantified by spectral EEG correlates in the low frequency range (1-7 Hz) during sleep and during wakefulness (Aeschbach, Matthews et al. 1997; Aeschbach, Matthews et al. 1999; Finelli, Baumann et al. 2000; Makeig, Jung et al. 2000; Cajochen, Wyatt et al. 2002).

Diverse predictive mathematical models have been developed to simulate process S (Borbely and Achermann 1992; Achermann, Dijk et al. 1993; Beersma 1998). Little is known about the brain structures involved in process S, and it seems that not only one but several neural correlates for S exist (Borbely and Tobler 1989; Harris 2005), such as adenosine activity from neuroglia cells (Jones 2009) that increase during prolonged wakefulness (Porkka-Heiskanen, Strecker et al. 1997; Latini and Pedata 2001). This could explain the sleep-inducing effect of SD. Caffeine has an antagonistic effect on adenosine receptors. In a recent study it could be shown that caffeine reduces the power density in the lowest delta band (Landolt, Dijk et al. 1995). This corroborates the potential major role of adenosine in the sleep homeostatic process.

The circadian System

On the other hand, **process C**, the circadian process, oscillates with a period of about 24 hours. It represents a clock-like process independent of whether one is asleep or awake and needs to be synchronized with external time (i.e. time of day) by so-called zeitgeber (i.e. light). Circadian rhythms are present at molecular, cellular, systems and behavioural levels. Molecular and cellular rhythms are found in nearly all cells and tissues in mammals (Buijs, Scheer et al. 2006; Kalsbeek, Palm et al. 2006; Maywood, O'Neill et al. 2006). There is almost no physiological or behavioural parameter that does not show circadian oscillations. Most physiological and behavioural parameters in humans, such as core body temperature (CBT), hormone levels (melatonin and cortisol), cognitive performance, subjective sleepiness and alertness and the sleep-wake rhythm, to mention only a few, undergo circadian rhythms with an approximate 24-h periodicity. The circadian process of C becomes evident under free-run conditions when external zeitgeber are absent (Aschoff 1965) (Webb and Agnew 1975). The circadian process is controlled by a concrete brain region, the first such pacemaker to be identified was the suprachiasmatic nuclei (SCN) located in the anterior hypothalamus. How exactly process C exerts its influence on process S is still unresolved. The entrainment pathway of the circadian timing system was shown to be a direct projection from the retina to the SCN via the retinohypothalamic tract (Hendrickson, Wagoner et al. 1972; Moore and Lenn 1972; Moore 1973) and ends as output of the sleep wake cycle, motor activity, mood, behaviour, hormonal secretion and neurobehavioral performance illustrated in Figure 1.2.

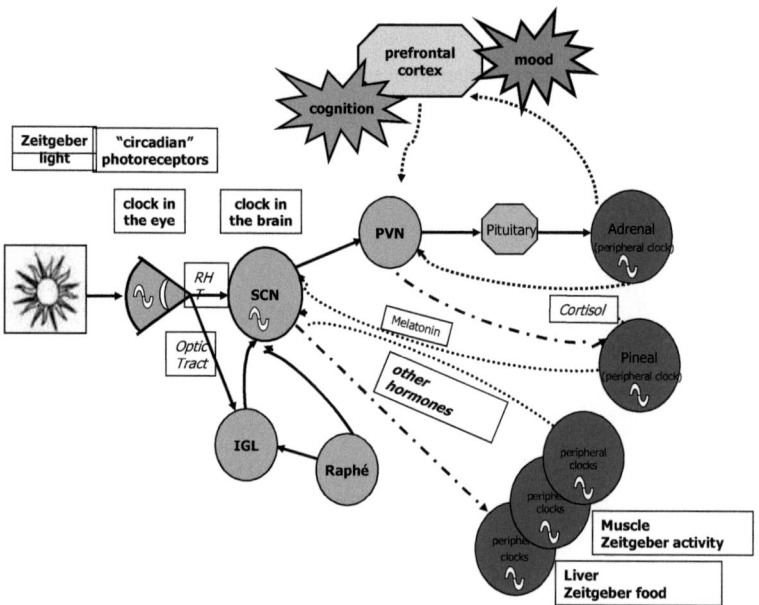

Figure 1.2: Schematic representation of the circadian system. The central pacemaker in the suprachiasmatic nuclei (SCN) receives light directly through retinohypothalamic tract (RHT). Peripheral clocks are coordinated by the SCN and are synchronized by specific zeitgeber (i.e. food, activity) (Wirz-Justice unpublished figure).

The characteristic of these rhythms may be significantly altered by the environment. This effect of exogenous and/or internal influences that modify the circadian rhythmicity independent of the biological clock is called masking.

There is growing evidence that genetic changes (Franken, Chollet et al. 2001; Retey, Adam et al. 2005) are also involved. In humans, genetic analysis has shown that clock gene variations are involved in the development of certain types of circadian rhythm sleep disorders, such as the advanced and delayed sleep phase syndrome (Pace-Schott and Hobson 2002; Ebisawa 2007).

Sleep structure has been shown to exhibit circadian regulation (Dijk and Czeisler 1995). The timing of REM sleep is strongly linked to the circadian rhythm, closely mirroring the core temperature. Thus, the maximum propensity for REM sleep is usually after the nadir of core temperature and it is less likely to occur during an afternoon and evening nap (Borbely 1982) (Dijk and Czeisler 1995). Furthermore, wakefulness during sleep or sleep continuity also depends strongly on the circadian pacemaker (Dijk and Czeisler 1995), such that sleeping outside the biological night (i.e. time of active melatonin secretion) results in an average sleep efficiency of only 70% outside the "melatonin window", compared to 95% within the "melatonin window" in normal healthy sleepers. Furthermore, wakefulness depends upon the maintenance of an activated cerebral cortex, which is accomplished through two mechanisms: input from multiple activating systems with critical input arising from the ARAS and, as we will note in a subsequent section, the opposition of homeostatic sleep drive by circadian drive for arousal (Moore 2007).

Interaction of homeostatic and circadian processes

These two distinct processes (homeostatic and circadian) interact to promote the onset of sleep when both are high (at the usual bedtime), and maintain sleep when process C is high and while process S is declining (in the early hours of morning). The so-called process S offsets deviations of sleep by augmenting sleep propensity when the duration of sleep has been curtailed, and by reducing sleep propensity in response to excessive sleep. Process C is considered to represent a wake-promoting drive to balance the homeostatic accumulation of sleep pressure illustrated in Figure 1.3.

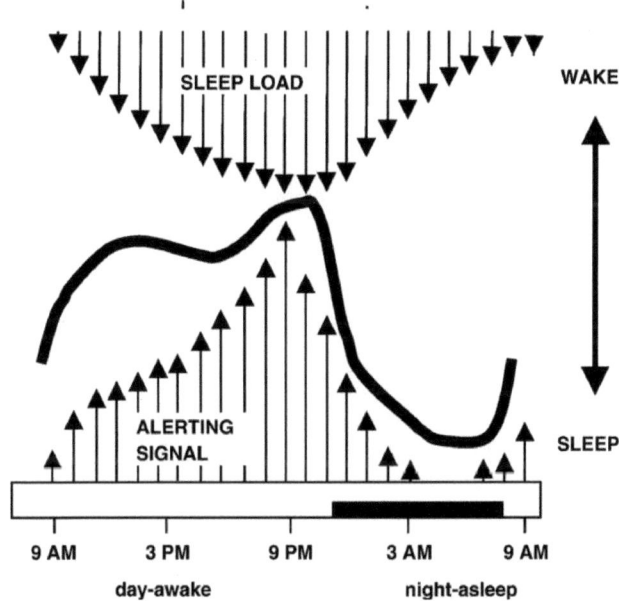

Figure 1.3: Schematic figure of the "opposing processes" mediating physiological sleepiness as a function of time of day. Sleep load increases in response to wakefulness imposed and/or maintained by the pacemaker

in the SCN. *Increasing levels of SCN-dependent alerting signals over the subjective day opposes homeostatic sleep drive, both of which peak shortly before the habitual sleep phase. Redrawn with permission from Edgar DM, Dement WC, Fuller CA. Effect of SCN lesions on sleep in squirrel monkeys: evidence for opponent processes in sleep–wake regulation. (Edgar, Dement et al. 1993)*

Process S and Process C and their interaction on selected physiological and neurobehavioral parameters

Much behaviour, like subjective mood, shows an underlying circadian rhythm modulated by the duration of prior wakefulness (Boivin, Czeisler et al. 1997) illustrated in Figure 1.4.

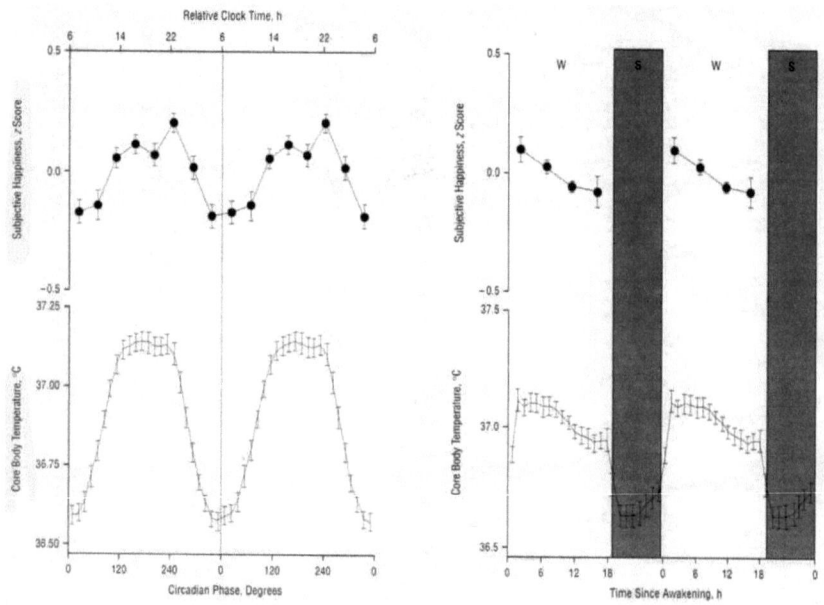

Figure 1.4: Circadian and wake-dependent variations of mood and educed waveform of core body temperature. Data from a forced desynchrony study. Mean±SEM values are plotted against circadian phase (left) or time since waking (right). W indicates wake, S asleep (Boivin, Czeisler et al. 1997).

The hormone **melatonin** serves as a chemical messenger of darkness in all species studied to date, including birds, fish, reptiles, insects, and mammals, and is an important component of the timing systems for circadian, and possibly infradian rhythms (Cassone and Natesan 1997). As such, the pattern of secretion of this chemical signal of darkness helps to induce night time behaviours. Melatonin also plays an important role in the modulation of circadian rhythms. Daily rhythmicity of melatonin secretion has been documented in different vertebrates and also in humans e.g. (Voultsios, Kennaway et al. 1997; Wright, Hughes et al. 2001). Melatonin does not depend on external stimuli (called zeitgeber) and persists in its absence; therefore the circadian rhythm of melatonin is not a result of masking.

Cortisol is a hormone secreted by the cortex of the adrenal gland. It plays an important role in the metabolism of glucose and proteins and has anti-inflammatory properties. Unlike the secretion of melatonin, the secretion of cortisol is not suppressed by photic stimulation; it is affected by various internal and external factors, the best known of which is stress. In humans, circadian rhythmicity in cortisol secretion has been documented by (Leproult, Van Reeth et al. 1997; van Eekelen, Kerkhof et al. 2003).

Subjective Sleepiness may be deemed as a physiological state or urge, that promotes the onset of sleep, and that is reversed or satiated by attainment of adequate sleep. The neuronal substrates of sleepiness are not understood completely. Sleepiness may reflect the waning of processes maintaining wakefulness or it may result from distinct neuronal systems acting to promote sleep (McCarley 2007). There exist different definitions of sleepiness, which is sometimes confused with fatigue, which is a very well known symptom of depression. Subjective sleepiness characterizes a need for sleep and is associated with an elevated disposition of falling asleep, usually known to disappear in stressful situations. Sleep propensity is often used as a synonym of sleepiness (Carskadon and Dement 1982; Johns 1998). Fatigue is more a subjective sensation of feeling tired especially in psychic, cognitive and somatic requirements.

Cognitive performance capability at any point in time is largely dependent on the two interacting homeostatic and circadian process systems, modulated by both C and S processes. The interaction of the two processes generates a constant vigilance performance throughout the entire day with 16-h duration. This opponent process is illustrated in figure 1.3. The homeostatic drive for sleep escalates with increasing time awake. The circadian system produces oscillations in functioning, which have an approximate frequency of one cycle every 24 hours; there does not exist an adequate theory to describe the underlying brain mechanisms responsible for these neurobehavioral deficits. With increasing time awake, particularly beyond the habitual amount of time spent awake, performance degrades considerably. Interestingly, to recover optimally from sleep loss-related performance decrements, less sleep is actually needed than the absolute hours of sleep formerly lost (Bonnet 2000). The circadian modulation of cognitive performance indicates a close temporal association with the circadian rhythm core body temperature with its maximum in the evening and nadir in the early morning (Rogers, Dorrian et al. 2003).

Chronobiology and depression – is there a link?

The concept of **depression** as a disease goes back a long way. In earlier times, depression was described as melancholia by many famous physicians. During the last century, psychiatric classification has been characterized by an inflation of diagnostic categories, including numerous subtypes of depression (see the numerous DSM and ICD classification systems). Severity, duration and recurrence are used as bases for classification. In terms of classification, depressive disorders, including the sub-classifications have long been recognized as heterogeneous. The core symptoms of MDD, of which at least one is required in DSM-IV, are depressed mood and loss of interest or pleasure. Further eligible symptoms are significant weight loss or gain, insomnia or hypersomnia, psychomotor agitation or retardation, fatigue or loss of energy, feelings of worthlessness or excessive or inappropriate guilt, diminished ability to think or concentrate, or indecisiveness, recurrent thoughts of death, recurrent suicidal ideation without a specific plan, suicide attempt or a specific plan for committing suicide. These symptoms have to occur almost daily during the same two week period within the most recent four week period, and represent a change from previous functioning of the person.

These core symptoms reflect the view that the depressive disorder is essentially a disorder of mood or affect and belongs as a subgroup to the mood respective affective disorders. Depression is a heterogeneous syndrome rather than a single disease, and has been characterized as a collection of "physiological, neuroencocrine, behavioural and psychological symptoms" (Nestler, Barrot et al. 2002; Fuchs, Simon et al. 2006).

According to the World Health Organization, depression affects nearly 121 million people worldwide. It is also one of the top 10 causes of morbidity and mortality (Rosenzweig-Lipson, Beyer et al. 2007).

Depression is a very complex disease and many gaps remain in our understanding of the aetiology and pathophysiology of depression. However, an increasing amount of evidence is now pointing to the possibility that chronobiological difficulties may underlie or at least accompany the condition. Numerous studies undertaken in depressive patients have suggested that a common comorbidity of depression is the dysregulation of the circadian timing system. This seems to occur in various types of depression, but is particularly evident in bipolar disorder (Bunney, Murphy et al. 1970; Wehr and Goodwin 1981). A particular feature of MDD is the abnormality in the timing and distribution of rapid eye movement (REM) and NREM sleep stages that can be regarded as a primary characteristic of the disease (Armitage 2007). Studies of sleep in depressive and affective disorders have been useful in supporting theoretical considerations about their pathophysiology (Srinivasan, Smits et al. 2006).

As we know from different contexts, most depressed suffer from insomnia. Paradoxically, it is hypothesized that according to the two process models of sleep regulation, SD may ameliorate the disease pattern, especially improving mood. Borbély and Wirz-Justice (Borbely and Wirz-Justice 1982) consider in their sleep regulation model that the depressed show a deficit in process S, which could manifest itself in slower build-up of sleep pressure during wakefulness or a different rate of decline during sleep (Figure 1.5 right). Hence, a full night of SD would raise it nearer to the normal values (Figure 1.5 right, dotted line). SD is conducive as a rapid and effective short-term treatment for depression (Wirz-Justice and Van den Hoofdakker 1999) (Giedke, Wormstall et al. 1990). A deficit in process C could be manifested in changed amplitude, phase or endogenous period (Figure 1.5 left). A shift in phase relationship between C and S can cause decrements in mood (Wirz-Justice 1995; Wirz-Justice 2003; Wirz-Justice 2006). In this context it has to be mentioned that it is thought that light therapy works by shifting the circadian rhythm (Wirz-Justice, Benedetti et al. 2005).

From pharmaceutical studies we have evidence that functional alterations in brainstem noradrenergic or serotonergic systems regulate mood and affective behaviours and the sleep-wake cycle. Imaging studies (Drevets 2001; Mayberg 2003) have since shown that several brain areas such as the prefrontal cortex, cingulated cortex, hippocampus, striatum, amygdale and thalamus are active during the experience of sleep disorders (insomnia and fatigue) as well as during disorders of mood (including depressed mood, feelings of worthlessness, diminished ability to concentrate, recurrent thoughts of death or suicide, as well as other symptoms as per the DSM-IV criteria for depressive disorders (American Psychiatric Association, 1994) (Association 1994)). Other studies have shown that an increased incidence of depressive symptoms correlates with poor sleep quality or chronic insomnia; disturbances that appear to be major risk factors for depression (Lustberg and Reynolds 2000).

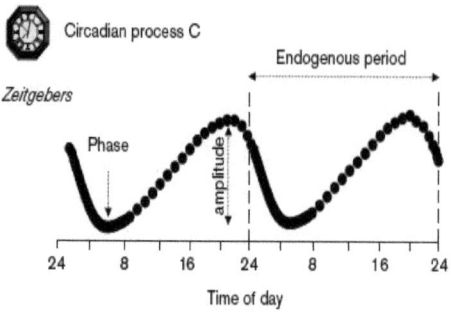

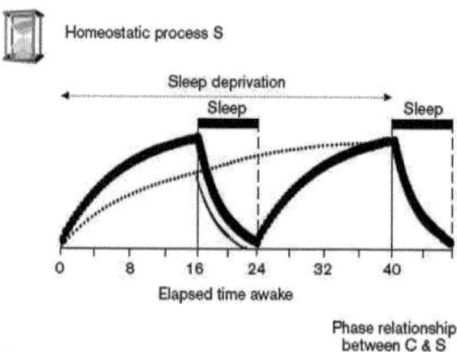

Figure 1.5: Schematic representation of the two-process models of sleep regulation. Above illustration shows the circadian process "C" which follows a 24-h rhythm independent of duration of prior wakefulness. In depression, it has been hypothesized that the circadian system might have an altered endogenous periodicity, advanced or delayed phase, and/or diminished amplitude. Additionally, the phase relationship between circadian and homeostatic processes could be abnormal. The illustration bellow depicts the homeostatic Process "S", which builds up exponentially during time awake and declines exponentially during sleep. The dotted line symbolises a putative disturbance in depression with a lower build up rate that only attains normality after 40-h of total SD (Wirz-Justice 2006).

Objectives and Methods

An important factor in circadian research is the so-called "masking" influence. Any external (e.g. light, body posture, food intake) or internal factor (e.g. stress level, motivation) has the potential to mask the "true" endogenous circadian rhythm. Therefore, these masking effects have to be controlled for. The constant routine protocol (CR) is a setting that realises minimalized and constant adherence of all known relevant masking factors. The CR is the gold standard to distinguish endogenous from exogenous drives of circadian rhythmicity. Study participants are kept under constant dim light levels, in semi-recumbent posture in bed, with isocaloric food and liquid intake spread over 24-h at regular intervals, and no time cues (Mills, Minors et al. 1978; Czeisler, Allan et al. 1986). The CR protocol in this thesis comprised 40-h of sustained wakefulness (the high sleep pressure protocol) and was contrasted against a 40-h period under CR conditions of sleep satiation realised by ten ultradian cycles of 225 minutes (150 min of scheduled wakefulness and 75 min of scheduled sleep; low sleep pressure protocol or nap protocol) demonstrated in figure 1.6. A limitation of the CR protocol is that there is no desynchronization between the sleep-wake cycle and the circadian pacemaker. Therefore, it does not allow for a segregation of these two processes. Nonetheless, investigation of the contribution of sleep pressure remains possible to some extent by comparing one condition in which subjects are totally sleep deprived during the entire experiment (high sleep pressure) to another condition in which they are allowed to nap regularly, the latter keeping homeostatic sleep pressure at a low level (low sleep pressure condition). Another very demanding protocol is the forced desynchrony protocol, which effectively uncouples the sleep-wake cycle and the endogenous circadian cycle. Both experimental paradigms expose the interaction of the homeostatic and circadian processes regulating sleep and wakefulness.

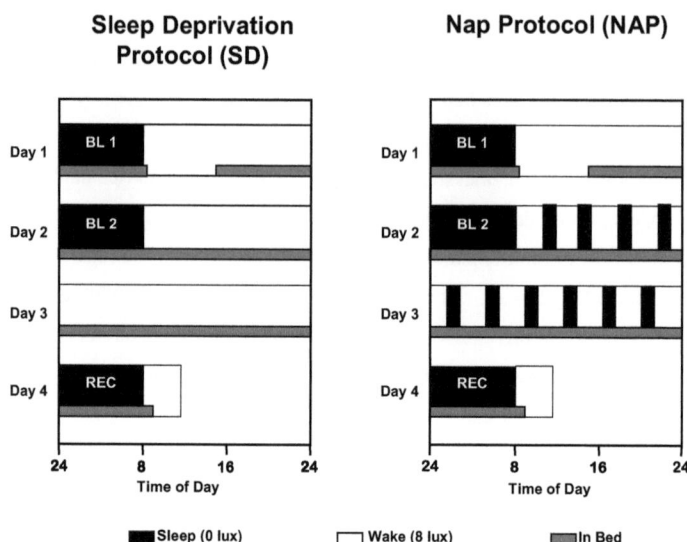

Figure 1.6: Schematic representation of the two 3.5-day protocols for high (left side) and low (right side) sleep pressure condition. White areas indicate scheduled wakefulness and dark bars delineate the scheduled sleep episodes. Grey bars depict constant posture, semi-recumbent during wakefulness and recumbent during sleep (without baseline week and adaptation night). BL=baseline night, REC=recovery night.

Outline of the thesis

The project on which the present thesis emerges tested the hypothesis whether major depression is a "clock"- or sleep disturbance or a "masking"-problem. In particular, the relative contributions of circadian and sleep homeostatic processes are investigated by carefully controlled protocols designed to minimize masking. Thus the following thesis aimed at elucidating the influence of the circadian system and the sleep–wake homeostat on mood, sleepiness, psychomotor vigilance performance, melatonin and the EEG during wakefulness in young women suffering from MDD in comparison to healthy age-matched women. Eight young women (24±4.8y) with MDD spent four days in the chronobiological facility at the Psychiatric University Clinics Basel and undertook the CR protocol during 40-h of sustained wakefulness or 40-h nap protocol attained with multiple naps. Depressed subjects selected for this study fulfilled the DSM-IV-TR diagnostic criteria for MDD without co-morbidity of another DSM-IV-TR disorder (axis I) and were in a major depressive episode undergoing the protocol. Equally, we have excluded depressed women with a high score (≥ 8) in the Pittsburgh sleep quality index (PSQI) (Buysse, Reynolds et al. 1989).

Subjective well-being, subjective sleepiness, melatonin and cortisol levels were first compared between two different age and gender groups of healthy participants under unmasked conditions in a high (i.e. SD) and low (i.e. Nap protocol) sleep pressure protocol of 64-h. These data served as the basic model to quantify circadian and sleep-wake related changes in subjective variables in healthy volunteers (Chapter 2). Before attempting to understand psychopathology it is important to consider *normal* daily variations of mood. The time course of subjective well being under constant routine conditions shows a circadian rhythm under low and high sleep pressure conditions (allegorized in Chapter 2).

In Chapter 3, the S-deficiency hypothesis of depression is tested by analysing waking EEG activity (an objective measure of sleepiness) and comparing the results against subjective sleepiness and tension during sustained wakefulness, using melatonin as the circadian marker of internal time. In Chapter 4, the focus is diurnal variations of mood and whether SD has an antidepressant effect. In the final Chapter 5, the time course of alertness in the depressive cohort as measured by a psychomotor vigilance performance test is investigated. Chapter 6 concludes by discussing the general implications of the findings for understanding putative circadian rhythm and sleep disturbances in depression, which may explain the efficacy of non-pharmalogical chronotherapeutics such as light therapy and SD.

In chapter 3, 4 and 5, data gathered from the same 64-h high and low-sleep pressure protocol from MDD patients are reported. The hypothesis that depression could be linked to a deficiency in the sleep-wake homeostatic process (i.e. the so-called S-deficiency hypothesis) was tested by waking EEG recordings during sustained wakefulness along with assessment of subjective sleepiness, tension and salivary melatonin (Chapter 3). Chapter 4 focussed on the investigation of diurnal variations in mood ratings and the well known antidepressant effect of SD in our cohort of MDD women. In a final step, in Chapter 5, psychomotor vigilance performance was analysed in MDD women under high and low sleep pressure conditions and compared against the healthy women. The general implications of these findings for understanding putative circadian rhythm disturbances in depression and the possibilities for forming a basis for the development of new treatment strategies such as light therapy and administration of chronotherapeutics (i.e. SD) are discussed in chapter 6.

Aims and hypotheses of the studies

The aim of this thesis was to investigate the influence of homeostatic and circadian processes in young depressive women by using the constant routine protocol under two differential sleep pressure conditions.

The specific goals in this project (Chapter 2-5) were:

A. **To investigate how subjective well-being, subjective sleepiness, salivary melatonin and salivary cortisol are modulated under two different sleep pressure conditions in healthy subjects, and to test if there are age and gender-related differences.**

Working Hypotheses: We expected that...

1. Subjective well-being is worse under high (SD) than under low sleep pressure conditions (nap protocol).
2. The circadian modulation of subjective well-being is more apparent in young than older study participants.
3. Women exhibit a more pronounced circadian modulation in subjective well-being than men.
4. Older study participants show lower subjective well-being ratings than young, particularly under high sleep pressure conditions.
5. The circadian modulation of subjective well-being exhibits a temporal correlation with other variables such as subjective sleepiness, cortisol, and the major circadian marker melatonin.

B. **To find out whether the very often proclaimed hypothesis of a deficiency in the homeostatic process of the sleep-wake regulation and diminished amplitude in circadian parameters, such as melatonin in depression, is verified, as well as a reduced response in frontal brain regions under sleep deprivation. And whether there is a correlation between subjective sleepiness and objective sleepiness measured by waking EEG.**

Working Hypotheses: We expected that...
1. Depressed women experience a deregulation of sleep-wake homeostasis in comparison to healthy women, as indexed by an altered time course in frontal low-EEG activity (1-7 Hz) during 40-h of sustained wakefulness.
2. Based on the deregulation of sleep-wake homeostasis, depressed women experience higher subjective sleepiness during 40-h of sustained wakefulness.
3. Depressed women show a diminished amplitude and/or circadian phase delay in salivary melatonin secretion under high sleep pressure conditions.

C. To affirm that sleep deprivation (high sleep pressure condition) improves mood in Major Depressive Disorder

Working Hypotheses: We expected that...
1. Depressive patients exhibit an improvement in subjective mood under high sleep pressure conditions compared to controls.
2. Under low sleep pressure, subjective mood displays a more circadian modulation than under high sleep pressure particularly in depressive women.

D. To investigate what role psychomotor vigilance plays in young women with major depression under different sleep pressure conditions? Does sleepiness differ from alertness? Are these antonyms? Does neurobehavioral performance demonstrate performance enhancement under the influence of the antidepressive effect of mood?

Working Hypotheses: We expected that...
1. Depressed women show worse psychomotor vigilance performance than controls. However, this difference is negligible with accumulating sleep pressure in the high sleep pressure protocol, due to the higher "prefrontal tiredness".
2. Psychomotor vigilance performance under low sleep pressure will be constantly reduced in depressed women compared to controls.
3. The improvement in psychomotor vigilance performance during high sleep pressure conditions in depressed women correlates with the antidepressant effect of SD.
4. Depressed women will undergo inferior physical comfort than healthy controls during the PVT performance under high sleep pressure conditions.

CHAPTER 2

Subjective Well-Being Is Modulated by Circadian Phase, Sleep Pressure, Age, and Gender

Angelina Birchler Pedross, Carmen M. Schröder, Mirjam Münch, Vera Knoblauch, Katharina Blatter, Corina Schnitzler Sack, Anna Wirz-Justice and Christian Cajochen

Centre for Chronobiology, Psychiatric Hospital of the University of Basel, CH-4025 Basel, Switzerland

Published in "Journal of Biological Rhythms" in June 2009
(J Biol Rhythms (2009); 24; 232-242)

ABSTRACT

Subjective well-being largely depends on mood, which shows circadian rhythmicity and can be linked to rhythms in many physiological circadian markers, such as melatonin and cortisol. In healthy young volunteers mood is influenced by an interaction of circadian phase and the duration of time awake. We analyzed this interaction under differential sleep pressure conditions in order to investigate age- and gender effects on subjective well-being. Sixteen healthy young (8 women, 8 men; 20-35 years) and 16 older volunteers (8 women, 8 men; 55-75 years) underwent a 40-h sleep deprivation (high sleep pressure) and a 40-h nap protocol (low sleep pressure) in a balanced cross-over design under constant routine conditions. Mood, tension and physical comfort were assessed by visual analogue scales during scheduled wakefulness, and averaged formed a composite score of well-being. Significant variations in well-being were determined by the factors "age", "sleep pressure" and "circadian phase". Well-being was generally worse under high than low sleep pressure. Older volunteers felt significantly worse than the young under both experimental conditions. Significant interactions were found between "sleep pressure" and "age", and between "sleep pressure" and "gender". This indicated that older volunteers and women responded with a greater impairment in well-being under high compared to low sleep pressure. The time course of well-being displayed a significant circadian modulation, particularly in women under high sleep pressure conditions. Our results demonstrate age- and/or gender-related modifications of well-being related to sleep deprivation and circadian phase and thus point to specific biological components of mood vulnerability.

Keywords: Mood, constant routine, sleep deprivation, sleepiness, melatonin, cortisol, sleep-wake homeostat, circadian rhythm

INTRODUCTION
Circadian and homeostatic influences

Subjective well-being largely depends on current mood, which is determined by both psychological and physical state. Under controlled laboratory conditions, subjective mood, assessed by a visual analogue scale (VAS), exhibits circadian rhythmicity (Boivin, Czeisler et al. 1997; Koorengevel, Beersma et al. 2003) similar to that of subjective sleepiness and cognitive performance (Van Dongen and Dinges 2005) and can be linked to rhythms in many physiological circadian markers such as core body temperature, heart rate, or the hormones melatonin and cortisol. In addition to the circadian component, manipulations of the sleep-wake cycle also have a strong impact on mood regulation such that subjective mood dramatically changed by altering the duration and timing of sleep episodes, thus suggesting that the duration of sleep and its position in the circadian cycle is critical for mood regulation (Taub and Berger 1973; Taub and Berger 1974; Monk, Buysse et al. 1992; Wood and Magnello 1992). Lack of sleep per se leads to mood deterioration in healthy subjects (Brendel, Reynolds et al. 1990; Scott, McNaughton et al. 2006) to such an extent that the timing of sleep, in sleep displacement studies, can significantly impact daily mean values of mood (Totterdell, Reynolds et al. 1994). Taken together, both circadian phase and the amount of prior wakefulness play a key role in the regulation of subjective mood.

Quantification of the differential influence of these two important factors in a forced desynchrony study design has revealed that subjective mood is modulated by a complex and non-additive interaction of circadian phase and duration of prior wakefulness (Boivin, Czeisler et al. 1997). The nature of this interaction was such that even moderate changes in the timing of the sleep-wake cycle led to profound effects on mood (Boivin, Czeisler et al. 1997). Similarly, after advancing the sleep-wake cycle daily by 20 min for a week, mood ratings fell strongly during the biological night compared with the stable sleep control group (Danilenko, Cajochen et al. 2003).

The Profile of Mood States questionnaire (POMS) (McNair, Lorr et al. 1971), a self-report inventory, is commonly used for measuring distinct mood states. Another frequently used instrument is the PANAS that measures two broad dimensions of Positive Affect (PA) and Negative Affect (NA) (Watson and Clark 1997). NA did not exhibit a clear circadian component whereas PA did and its 24-hour rhythm correlated with the circadian rhythm of rectal temperature (Murray, Allen et al. 2002). Affective state, as measured by various mood and subjective activation scales (e.g. (Covi, Lipman et al. 1977; Monk 1989)), is also sensitive to sleep loss (Reynolds, Kupfer et al. 1986; Bonnet 1989). Another interesting aspect is that the characteristics of morningness (M) or eveningness (E) in healthy subjects significantly impact on their mood states (measured by POMS):

E chronotypes improved mood and decreased anger-hostility after partial and total sleep deprivation, and activity was decreased after total sleep deprivation. For M chronotypes partial sleep deprivation did not modify mood, whereas total sleep deprivation worsened depressive mood and tiredness, and decreased vigor-activity (Selvi, Gulec et al. 2007).

Age and gender related influences

Younger subjects consistently rated themselves lower on global measures of vigour and affect than older subjects, with a sharper decrease of vigour on the day following sleep deprivation (Brendel, Reynolds et al. 1990). This suggests that acute sleep deprivation may actually be more disruptive for younger than for older adults, who may have greater mood stability and less rhythmic changes (Monk, Buysse et al. 1992). However another study under 36-h bed rest conditions revealed no differences in temporal profiles between older and young volunteers (Buysse, Monk et al. 1993).

No significant gender effect on mood was found in a study with pilots (Caldwell and LeDuc 1998), although another study showed that the diurnal rhythm of mood in women peaked 2 hours earlier than men (Adan and Sanchez-Turet 2001). However, to our knowledge there are no studies on age and gender-related differences in circadian and homeostatic mood regulation so far.

Study aim

We investigated the time course of subjective mood, tension and physical comfort ratings in young and older healthy subjects under differential sleep pressure conditions in order to quantify circadian and homeostatic contributions to these ratings. Because the 40-hour protocol assesses subjective well-being at a high sampling frequency (in addition to collecting physiological variables and carrying out performance and memory tests), we could not implement a large questionnaire battery, but selected the previously validated, readily comprehensible and fast VAS technique to maintain subjects' motivation. A significant concern in the present study was that physical discomfort and tension arising in the course of the demanding protocol (64 hours of bed rest) may impact subjective mood ratings, particularly in the older cohort. Thus, we decided to combine tension and physical comfort together with subjective mood into a composite score and defined this score as an index of subjective well-being.

We utilised a very strictly controlled constant routine protocol to minimise the majority of confounding ("masking") factors [for details see (Cajochen, Knoblauch et al. 2001)].

Based on the fact that more women than men suffer from mood disorders and older subjects experience more physical problems and reduced circadian modulation than young that will manifest itself particularly under high sleep pressure conditions we predicted that:

1. Subjective well-being is worse under high (sleep deprivation) than under low sleep pressure conditions (nap protocol).
2. The circadian modulation of subjective well-being is more apparent in young than older study participants.
3. Women exhibit a more pronounced circadian modulation in subjective-well being than men.
4. Older study participants show lower subjective well-being ratings than young, particularly under high sleep pressure conditions.
5. The circadian modulation of subjective well-being exhibits a temporal correlation with other variables such as subjective sleepiness, cortisol and the major circadian marker melatonin.

METHODS
Study participants

All study participants were recruited via advertisements at different Swiss universities and in newspapers. Sixteen healthy young (8 women and 8 men, age range 20-31y, mean age 25.0±3.3y [SD]) and 16 healthy older volunteers (8 women and 8 men age range 57-74y, mean age 65.0±5.5y) were selected. The mean body mass index (BMI) was 21.5±1.6 [SD] for the young and 23.3±2.1 for the older volunteers (t-test: $p < 0.05$). Each study volunteer underwent a physical examination, an interview about sleep quality, life habits and health state, a neuropsychological test battery [CANTAB® test battery and the Stroop Test (only for the older group)]. All were free of medical, psychiatric, neurological and sleep disorders [as per Pittsburgh Sleep Quality Index (PSQI) score ≤5 (Buysse, Monk et al. 1993), and a polysomnographically (PSG) recorded screening night]. The mean PSQI value was 2.1±1.3 [SD] for the young and 3.4±1.7 S.D. (t-test: $p < 0.05$) for the older volunteers. Volunteers were included if their clinical sleep EEG scoring had no pathological findings [apnoea/hypopnoea-index (AHI) <10/h; periodic leg movements (PLM) index <10/h]. To exclude chronotype-specific differences in circadian phase preference we selected only moderate chronotypes (morning-evening-type [M/E] Questionnaire rating between 14 and 21 points) (Torsvall and Åkerstedt 1980). Nevertheless, M/E scores were slightly higher in the older than in the younger group (mean±SEM: 18.8±0.8 *vs.* 16.4±0.8; t test: $p<0.05$), corresponding to an earlier chronotype. All participants were non-smokers without any drug abuse. This was verified in the young group by urinary toxicological analysis, sensitive for amphetamines, benzodiazepines, opiates and tetrahydrocannabinol (Drug-Screen Card Multi- 6®, von Minden GmbH, Moers, Germany). Participants were also required to abstain from excessive caffeine and alcohol consumption as well as heavy physical exercise. Other exclusion criteria were: shift work within 3 months and transmeridian flights within 1 month prior to the study, excessive caffeine and alcohol. The young women started the study on days 1–5 after menses onset in order to complete the entire study block within the follicular phase, with the exception of five young women taking oral contraceptives.

All procedures conformed to the Declaration of Helsinki. The local Ethical Committee approved the study protocol, screening questionnaires and consent form [for details see (Munch, Knoblauch et al. 2004; Munch, Knoblauch et al. 2007)] and all study participants gave signed informed consent.

Protocol and study design

Each participant was instructed to maintain a regular sleep–wake cycle (bed- and wake-times within ±30 min of self-selected target time), which was verified by wrist activity monitors (Cambridge Neurotechnology®, UK) and sleep logs during one week prior to study begin. Habitual bedtimes did not vary significantly between groups (young: 23:34±56 min *vs.* older: 23:11±40 min; mean±SD; p=0.2, t-test). The study entailed a balanced and gender-matched cross-over design, each block lasting 3.5 days (details in Figure 2.1), and started with an 8-h PSG night in the laboratory. During day 1 subjects adjusted to the experimental dim light condition (<8 lux), and a morning a blood sample was taken from the older participants to verify both a normal haemogram and physiological coagulation; they received a low-dose heparin injection on the three consecutive days of each study block (Fragmin® 0.2 ml, 2500 IE/Ul, Pharmacia AG, Dübendorf, Switzerland) in order to prevent any venous thrombosis. After a second 8-h sleep episode, all subjects participated in a 40-h "constant routine" (CR) protocol as detailed in (Cajochen, Knoblauch et al. 2001; Munch, Knoblauch et al. 2004; Knoblauch, Munch et al. 2005; Munch, Knoblauch et al. 2007), which was followed by a recovery night.

The subjects underwent two CR conditions spaced 1-3 weeks apart: a high sleep pressure [40-h sleep deprivation (SD) protocol] and a low sleep pressure (10 cycles of 150 min scheduled wake: 75 min scheduled sleep; NAP protocol). The 8-h sleep episode was calculated with respect to the midpoint of each individual's habitual sleep episode as assessed by actigraphy and sleep logs during the baseline week. All wake episodes were spent under semi-recumbent CR conditions (<8 lux) during wakefulness with a minor shift to supine posture during scheduled sleep episodes (0 lux).

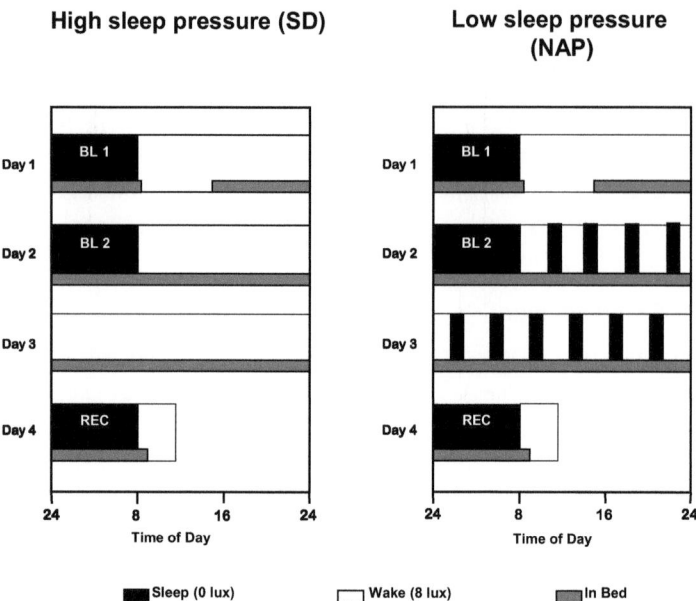

Figure 2.1: Overview of the 3.5-day laboratory protocol (without baseline week and adaptation night). Black bars (0 lux) indicate scheduled sleep episodes and white bars scheduled wake episodes (<8lux). Grey bars denote controlled posture (semi-recumbent during wakefulness and supine during sleep). BL=baseline night, REC=recovery night. The data analysed in this paper were gathered in the 40 hours between baseline night 2 and the recovery night, which was a constant routine protocol with either total sleep deprivation or multiple naps.

Subjective rating scales

Subjective well-being was a composite score averaged over the three items "mood, tension, and physical comfort", each assessed by a 100-mm bipolar VAS at 30-min intervals. The participants were asked to indicate how he or she felt "at that moment" by placing a vertical mark on the VAS ranging from 0 ("worst ever") to 100 mm ("best ever"). Since the direction of the extremes was not the same for all the three items the formula was as follows: *subjective well-being = [VAS$_{mood}$+ (100-VAS$_{tension}$) + VAS$_{physical\ comfort}$]/3*. In addition, VAS estimates of alertness, hunger, and subjective thermal comfort were collected. Although the reliability and validity of VAS especially in measuring emotions have been confirmed in many studies (Aitken 1969); (Folstein and Luria 1973), our index of subjective well-being has not yet been validated. However, we have first evidence from an ongoing constant routine study in our laboratory that the dimension of positive affects in the PANAS correlate rather well with the composite score of subjective well-being

($r=0.66$, $p<0.025$) and even more so for the specific items VAS_{mood} and the item *happiness* on the PANAS ($r=0.74$; $p<0.01$).

Subjective sleepiness was assessed by the composite score of the Karolinska Sleepiness Scale (KSS) and the Karolinska Sleepiness Symptoms Check List (KSSCL) at 30-min intervals.

Salivary assays

Saliva collections for hormonal assays were scheduled during wakefulness at the same 30 min time intervals as subjective ratings.

Melatonin: a direct double-antibody radioimmunoassay (RIA) was used for the melatonin assay (validated by gas chromatography–mass spectroscopy with an analytical least detectable dose of 0.65 pm/ml); Bühlmann Laboratories, Schönenbuch, Switzerland (Weber, JC et al. 1997).

For mean melatonin levels, values of all samples between the upward- and downward-mean crossing points were averaged per subject and age group. A nap was classified as occuring during the biological night if the melatonin concentration of the last saliva sample prior to the nap was above the individual mean, otherwise it was classified as a nap during the biological day (Knoblauch, Munch et al. 2005; Munch, Knoblauch et al. 2005).

Cortisol: Cortisol was measured by RIA (Ciba Corning Diagnostics) with a detection limit of 0.2 nmol/l. The intra-assay CV was 4.0% above 0.4 nmol/l and 10.0% for levels below.

Data analyses and statistics

For data reduction, all values were collapsed into 3.75-h bins per subject before averaging over subjects. For all analyses, the statistical packages SAS® (SAS Institute Inc., Cary, NC, USA; Version 6.12) and Statistica® (Stat-Soft Inc., 2000–2004, STATISTICA for Windows, and Tulsa, OK, USA) were used. Four-way repeated measures ANOVA (rANOVA) with the factors 'age' (young *vs.* older), 'gender' (women *vs.* men), and the repeated factors 'sleep pressure' (high *vs.* low sleep pressure condition) and 'time of day' (11 time points) were performed. All p-values derived from rANOVAs were based on Huynh-Feldt's (H-F) corrected degrees of freedom (significance level: $p < 0.05$). At some time points the data for different variables (e.g. subjective well-being, melatonin etc.) were not normally distributed, and thus a non-parametric test was used for post-hoc comparisons (Mann-Whitney U test). Backward stepwise regression analysis was performed to identify the important predictor variables among subjective sleepiness, cortisol and melatonin for subjective well-being.

RESULTS

Subjective well-being

Mean subjective well-being ratings in the course of the high (SD) and low sleep pressure (NAP) protocol for the young and older women and men are illustrated in Figures 2.2 and 2.3. In general, all participants assessed their subjective well-being as better than average with an initial score above 50 (0= worst ever and 100= best ever). The rANOVA yielded significance for the main factors 'age', 'sleep pressure' and 'time of day' (Table 1). Average well-being was significantly lower in older participants than the young (56.9±2.2 vs. 65.3±2.1), and lower during SD than NAP conditions (59.7±1.9 vs. 62.4±1.6). A circadian modulation revealed lower ratings during the biological night compared to the biological day. Well-being of older participants was more impaired under SD conditions than the young (-5.5±2.4 vs. -0.1±2.2; interaction 'age x sleep pressure', Table 2.1). The significant two-way interaction 'gender x sleep pressure' emphasizes a clear decrement in subjective well-being in women but not in men under the SD condition (-6.0±2.4 vs. 0.4±2.6). Young men did not seem to be affected by rising sleepiness under SD (Figure 2.3), and show a rather flat curve throughout. In contrast, subjective well-being declined in both young and older women during the evening (significant difference between young women and men between 17:30 and 04:30 h, p at least 0.04, Mann Whitney U-test). In the NAP protocol we only found a significant time of day effect. This protocol has the characteristic of revealing the underlying circadian rhythm since sleep pressure does not rise to mask it

Factor	Subjective Well-being	Subjective Sleepiness	Melatonin	Cortisol
Age group	P=0.012	P=0.048	P=0.03	n.s.
Gender	n.s.	P=0.053	n.s.	n.s.
Sleep Pressure	P=0.009	P<0.0001	P=0.03	n.s.
Time of Day	P<0.0002	P<0.0001	P<0.0001	P<0.0001
Age x Gender	n.s.	n.s.	n.s.	n.s.
Age x Sleep Pressure	P=0.012	n.s.	P=0.0037	n.s.
Age x Time of Day	n.s.	n.s.	n.s.	P=0.03
Age x Time of Day x Sleep Pressure	n.s.	n.s.	P=0.014	n.s.
Gender x Sleep Pressure	P=0.003	n.s.	n.s.	n.s.
Gender x Time of Day	n.s.	P=0.045	n.s.	P=0.001
Sleep Pressure x Time of Day	n.s.	P<0.0001	P=0.03	n.s.
Sleep Pressure x Age x Gender	n.s.	n.s.	n.s.	n.s.
Time of Day x Age x Gender	n.s.	n.s.	n.s.	n.s.

| Time of Day x Age x Gender x Sleep Pressure | n.s. | n.s. | n.s. | P=0.02 |

Table 2.1: Main and interaction effects of age, gender, sleep pressure and time-of-day on subjective well-being, subjective sleepiness, and salivary melatonin and cortisol levels. n.s. = non significant (p>0.05).

Subjective sleepiness

The time course of subjective sleepiness is the second panel in Figures 2.2 and 2.3. The rANOVA yielded significance for the main factors 'age', 'sleep pressure' and 'time of day' and was at near significance for the factor 'gender' (p=0.053; Table 2.1). Older volunteers were on average sleepier than the young (4.2±0.3 *vs.* 3.5±0.2), and all participants were sleepier during high than low sleep pressure conditions (4.9±0.3 *vs.* 2.8±0.1). Significant two-way interactions were found for 'gender' x 'time of day' as well as for 'sleep pressure' x 'time of day'. Women were sleepier than men particularly during the biological night and during the SD protocol. Young women were sleepier than young men in the SD protocol from 12:00 h the first day until 18:45 h the next day (*p* at least 0.04, Mann Whitney U-test). There was a marked circadian modulation of sleepiness in all sleep pressure conditions and age groups (previously reported for the nap protocol (Munch, Knoblauch et al. 2005)).

Melatonin

The time course of salivary melatonin concentration is illustrated in the third panel of figures 2.2 and 2.3. The rANOVA yielded significance for the main factors 'age', 'sleep pressure' and 'time of day' (Table 1). Older volunteers had significantly lower mean melatonin levels than the young (3.8±0.5 *vs.* 6.4±1.1 pg/ml), and all volunteers had slightly but significantly higher melatonin levels during the low (NAP) than the high sleep pressure (SD) protocol (5.4±0.7 *vs.* 4.8±0.6 pg/ml). The significant two-way interaction 'age' x 'sleep pressure' indicated that elevated melatonin levels under low compared to high sleep pressure conditions were only seen in the young but not the older volunteers. Furthermore, the significant two-way interaction 'sleep pressure' x 'time of day' and a significant three-way interaction 'age' x 'sleep pressure' x 'time of day' was observed. Post-hoc comparisons revealed significantly higher melatonin levels in young women compared to young men between 02:45-06:30 h (SD) during high sleep pressure (SD) and 17:00-10:00 h the next day (NAP) (*p* at least 0.04, Mann Whitney U-test).

Although melatonin secretion was diminished in older volunteers, there were no significant differences compared to the young in circadian phase position or timing of the sleep wake cycle, nor did the phase angle between them differ [results published in (Knoblauch, Munch et al. 2005; Munch, Knoblauch et al. 2005)].

Cortisol

The time course of salivary cortisol concentration is illustrated in the bottom panel of figures 2.2 and 2.3. Only the main factor 'time of day' yielded significance (Table 1). However, the four-way interaction ('age' x 'gender' x 'sleep pressure' x 'time of day') as well as the two-way interactions 'age' x 'time of day' and 'gender' x 'time of day' were significant. The former most likely reflects a reduced circadian profile in cortisol secretion in the older subjects. The latter reflects higher cortisol levels in the evening in young women than young men between 23:00-6:30 h, and lower cortisol levels from 14:00-17:45 h on the second day ($p < 0.05$ Mann Whitney U-test).

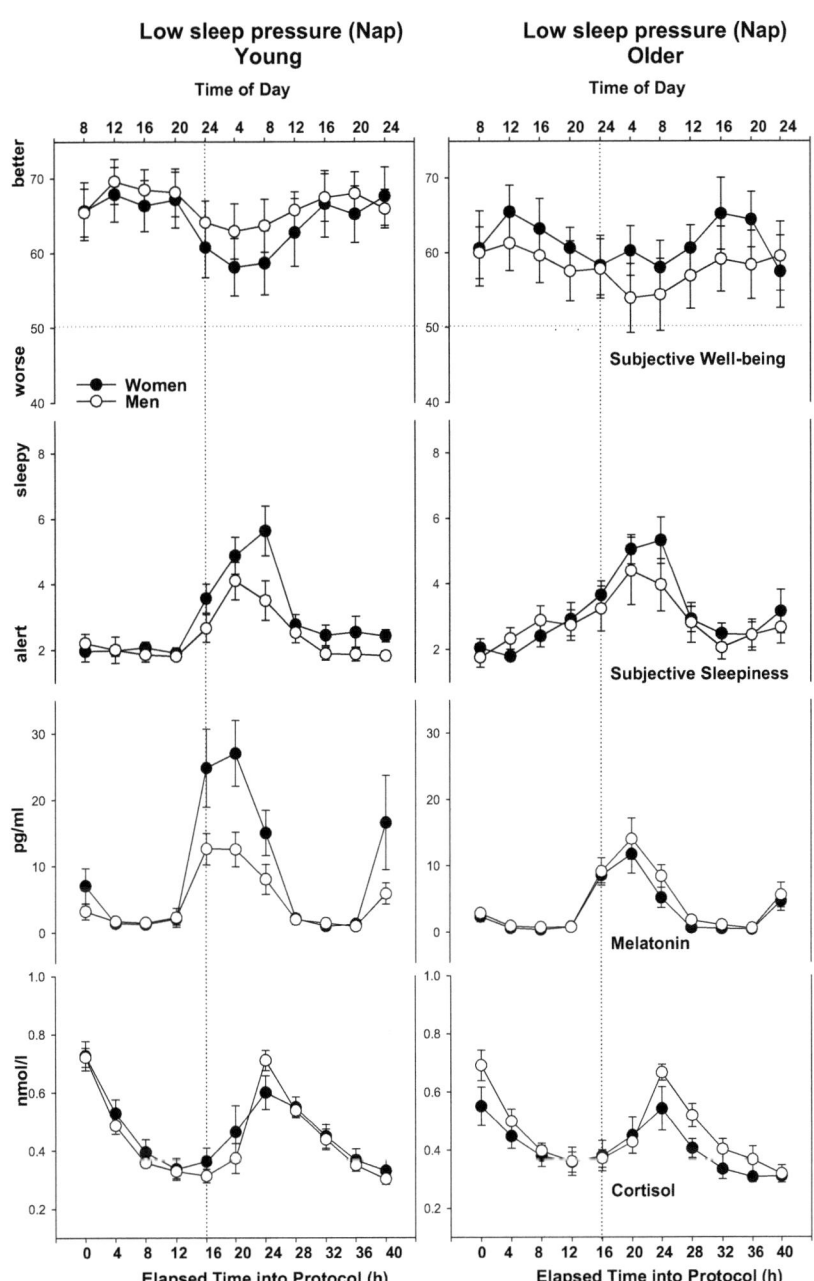

Figure 2.2: Time course of subjective well-being in young (left) and older (right) volunteers under low sleep pressure conditions (nap protocol; mean values per 3.75-h bin±SEM; n=16). Open circles represent men and filled black circles women (Top Panel). The x- axis above the figure describes time of day (in hours)

and the x-axis below describes elapsed time into protocol (time course over 40 hours). Subjective well-being is expressed as the average of three different 100mm visual analogue scales (mood, tension, and physical comfort). Panel 2: Time course of subjective sleepiness (composite score of KSS and KSSCL). Panel 3: Time course of salivary melatonin concentrations (pg/ml). Bottom Panel: Time course of salivary cortisol secretion (nmol/L).

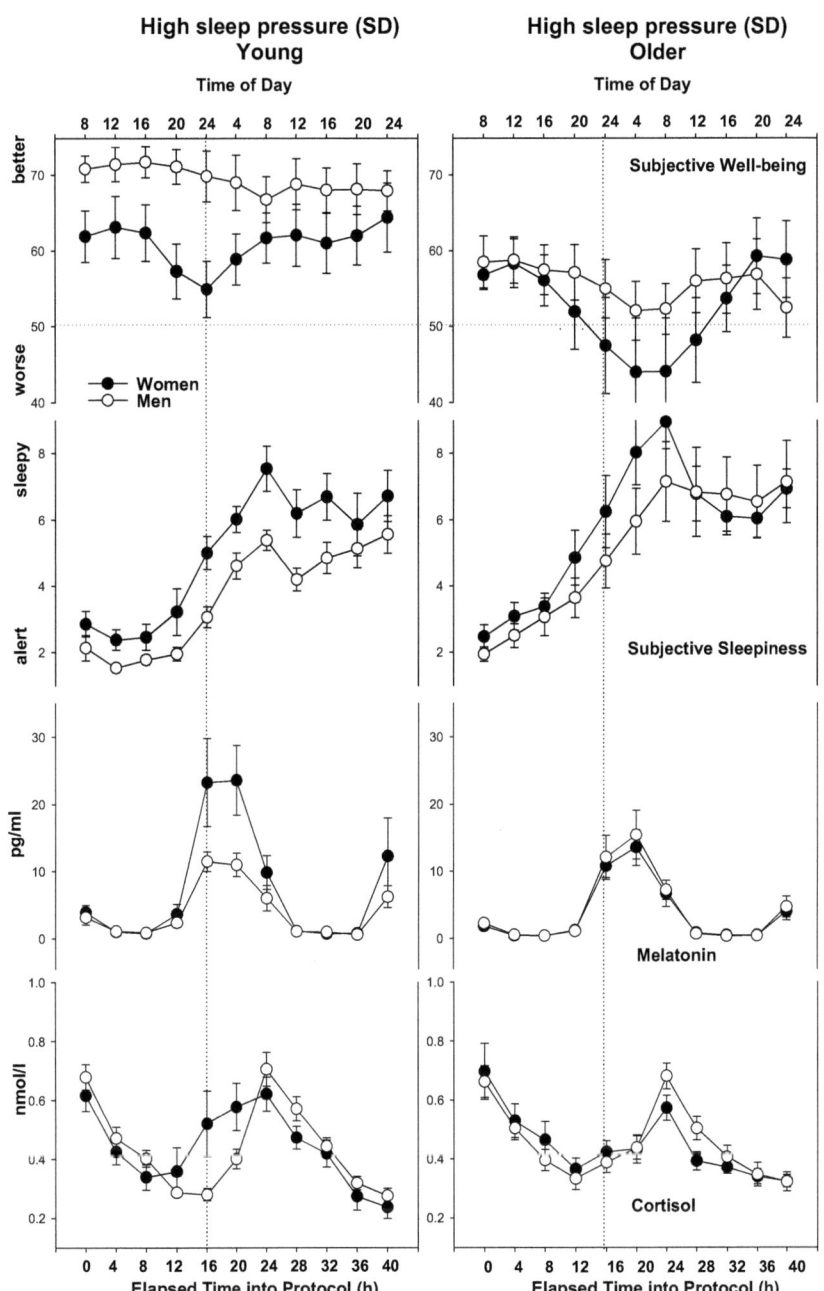

Figure 2.3: Time course of subjective well-being in young (left) and older (right) volunteers under high sleep pressure conditions (sleep deprivation protocol; mean values per 3.75-h bin±SEM; n=16) (Top Panel).

Details as in Figure 2. Panel 2: Time course of subjective sleepiness. Panel 3: Time course of salivary melatonin concentrations. Bottom Panel: Time course of salivary cortisol secretion.

Subjective sleepiness and cortisol levels as predictors for subjective well-being

In order to investigate possible relationships between subjective well-being, subjective sleepiness, and the circadian marker cortisol, a backward stepwise regression analysis was calculated (Table 2.2). Subjective well-being showed the highest correlations (r = -0.45) with subjective sleepiness followed by cortisol (r = - 0.15), while melatonin was excluded by the regression model. Thus, subjective sleepiness can explain in general about 20% of the variation in the subjective well-being ratings of our data pool. To further test whether this association depended on circadian phase and sleep pressure conditions, subjective well-being and sleepiness were correlated at the 11 different time points throughout the high and low sleep pressure protocol separately (Figure 2.4). For time points exceeding the usual 16 hours of wakefulness, we found a significant correlation between subjective well-being and sleepiness. Up to 50 percent of the variation in subjective well-being could be explained by subjective sleepiness at times when high sleep pressure coincided with the circadian trough (4 am). Under low sleep pressure conditions, however, there was never a significant correlation observed, despite the evidence for a clear circadian impact on both subjective well-being and sleepiness ratings.

Variable	df	F	r	P
Subjects	31			
Subjective sleepiness	1	180.5	-0.45	0.0001
Cortisol	1	20.2	-0.15	0.0008
Residuals	690			
Total	721			

Table 2.2: Results of the backward stepwise regression analysis.

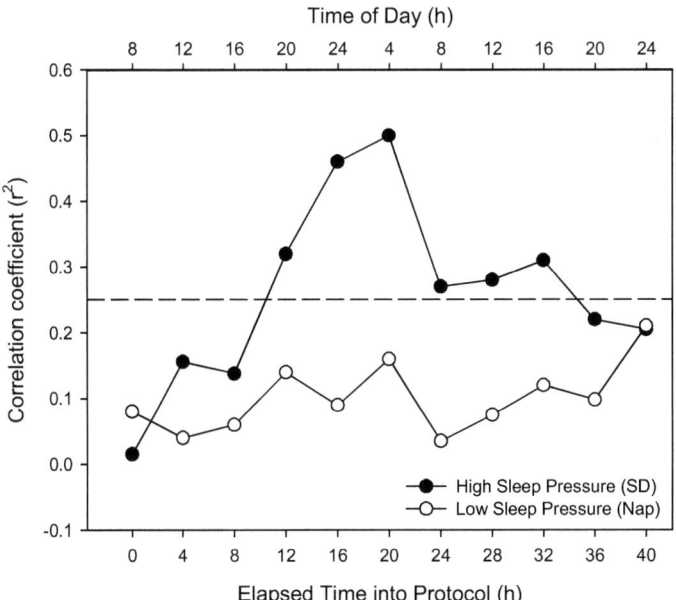

Figure 2.4: Time course of the correlation coefficient (r^2) between subjective well-being and subjective sleepiness in the course of the high (closed symbols) and low sleep pressure protocol (open symbols). The stippled horizontal line delineates the threshold between significant and nonsignificant correlations ($p<0.05$, $n=32$ for each time point).

DISCUSSION

Despite the very demanding constant posture conditions (64 hours in bed), our healthy study participants rated their subjective well-being in general as good and never attained low levels. As predicted, well-being was modulated by time of day (worse during the biological night than during the biological day) and by sleep pressure (worse during SD than NAP). This confirms an important role of the circadian and the sleep-wake homeostatic system on subjective well-being, modified by age and gender. In the following section we will discuss the impact of the different factors.

Circadian modulation and sleep pressure

The time of day modulation of subjective well-being was prominent in both protocols, indicating that circadian phase plays a pivotal role in well-being. This is in accordance with forced desynchrony data in which a significant circadian component of mood regulation could be educed (Boivin, Czeisler et al. 1997; Koorengevel, Beersma et al. 2003). Not clear is the role of wakefulness duration on mood, visible in our CR and in the short forced desynchrony protocol by Koorengevel et al., 2003 (Koorengevel, Beersma et al. 2003), but not in the classic study (Boivin,

Czeisler et al. 1997). All our participants in the SD protocol showed deterioration of subjective well-being following 24 hours of wakefulness, declining about 4 hours earlier in women than in men and improving again in young women already after 22 hours of prior wakefulness. Could women have had a higher motivation to overcome sleepiness? More plausible is the higher stress (see increased cortisol levels in the early evening, Figure 2.3) in our young female participants when they faced having to stay awake all night. The circadian component again was visible in the improved well-being that returned on day 2, despite very high sleepiness levels.

Are women more vulnerable to circadian and sleep-homeostatic influences, or do they "sense" these changes better than men? Are men less "sensitive", less attentive of their well-being, or are they more socially conditioned to not show negative emotions? Men had smaller circadian amplitude in well-being during high sleep pressure conditions, suggesting more stability. Indeed, well-being in men did not decline under high sleep pressure compared to low sleep pressure conditions, whereas it did in women. The near-linear pattern in young men was remarkable. Is this male insensitivity to circadian and sleep homeostatic alterations in well-being and sleepiness related to higher risk-taking behaviours in young men, particularly when sleep deprived (e.g. their greater accident rate (Horne and Reyner 1995; Pack, Pack et al. 1995)? Of course, there are major differences between real and simulated environments (Philip, Sagaspe et al. 2005), and cognitive performance is highly influenced by motivation (Hull, Wright et al. 2003).

Age-effects

Subjective well-being was significantly lower in our healthy older cohort than in the young, whether due to a decrease in subjective perception or diminishing physiological and/or psychological constitution. The demanding study protocol could have been experienced more negatively in the older group. Surprisingly, the older subjects did not show a reduced circadian modulation of well-being, as originally hypothesized, but a tendency to an even more prominent circadian rhythm in subjective well-being than young subjects.

However, the older group responded to high sleep pressure with a significant greater deterioration in well-being than the young. This implies a greater vulnerability to changes in circadian phase and challenges of sleep pressure with age. In contrast, other measures such as psychomotor vigilance performance appear less vulnerable to sleep debt in the older, not only in our study (Adam, Retey et al. 2006; Blatter, Graw et al. 2006). Even middle-aged subjects are less vulnerable than the young in this aspect of behaviour (Philip, Taillard et al. 2004; Bliese, Wesensten et al. 2006). Thus we could only partially confirm previous findings (Brendel, Reynolds et al. 1990) of greater mood and performance disturbances after sleep loss in older subjects.

Other circadian parameters

It is interesting to compare subjective well-being with other circadian parameters, such as subjective sleepiness, melatonin [data already partially published (Cajochen, Knoblauch et al. 2001; Knoblauch, Martens et al. 2003; Cajochen, Knoblauch et al. 2004; Knoblauch, Munch et al. 2005; Munch, Knoblauch et al. 2005; Cajochen, Munch et al. 2006)] and cortisol concentrations. Although our results indicate that subjective sleepiness and cortisol levels were significantly correlated with subjective well-being, these variables had dissimilar time courses under both protocols. Increasing sleepiness under high sleep pressure exhibited significant repercussions on subjective well-being in the early morning when increased homeostatic sleep load coincided with maximal circadian drive for sleep (corresponding time of day: 4 am, figure 2.4), but not when highest sleep pressure (38-40 hours) coincided with the maximal circadian drive for wakefulness in the evening (corresponding time of day: 10 pm -midnight the second day of the protocol, figure 2.4). In contrast, sleepiness under low sleep pressure manifested a circadian rhythm similar to subjective well-being, but no significant correlation between these measures were observed at any given time point. Thus, the relationship between subjective sleepiness and well-being is not trivial but depends on a complex interaction between the circadian pacemaker and the sleep homeostat.

The circadian melatonin and cortisol profiles followed the well-known temporal dynamics under CR conditions [reviewed in (Arendt 2006)]. The cortisol rhythm did not change with age, but nocturnal melatonin was significantly attenuated. We do not have any explanation for the lower nocturnal melatonin levels in young men – apart from possible chance differences in this particular group (there are much greater inter-individual differences in melatonin than there are differences between men and women (Arendt 2006)). The elevated cortisol levels around the evening nadir in young women may have reflected their more "stressful" reaction to the SD protocol, and this was correlated with well-being in a backward regression analysis. However, other studies have shown no or only minimal stimulation of cortisol secretion by sleep deprivation (Scheen, Byrne et al. 1996; Brun, Chamba et al. 1998).

Limitations of the study

Measuring behaviour under highly controlled laboratory conditions is important to assess contributions of circadian and homeostatic processes to a subjective variable such as well-being. However, the results are not directly applicable to understanding squeal of a chronic sleep deficit caused by sleep problems or shift work in real life. Chronic partial sleep restriction shows a different dynamic profile than acute total sleep deprivation (Banks and Dinges 2007), and a much smaller change in subjective ratings (Brunner, Dijk et al. 1993; Belenky, Wesensten et al. 2003).

CONCLUSION

Our results demonstrate clear age and gender-related modification of circadian and sleep-wake-homoeostatic contributions to subjective well-being. In general, both older adults and women were more affected by sleep deprivation, showing a tendency to lower subjective well-being and a prominent circadian trough. Given that circadian and sleep homeostatic processes regulate mood in healthy subjects, it is not surprising that the circadian dysregulation and sleep disturbances associated with depression may have profound detrimental effects on mood in depressed patients, thus further perpetuating the disorder.

ACKNOWLEDGMENTS

We are very grateful to our technicians Claudia Renz, Marie-France Dattler, Giovanni Balestrieri and the student shift workers for their help in data acquisition. We thank Gabrielle Brandenberger for coordinating the analysis of cortisol samples and Sarah Chellappa for her linguistic editing support. This research was supported by Swiss National Science Foundation Grants START # 3100-055385.98, and 3130-0544991.98 and 320000-108108 to CC, the Velux Foundation (Switzerland) and Bühlmann Laboratories, Allschwil (Switzerland).

CHAPTER 3

Higher Frontal EEG Synchronization in Young Women with Major Depression: A Marker for increased Homeostatic Sleep Pressure?

Angelina Birchler-Pedross, PhD[1]; Sylvia Frey, PhD [1]; Sarah Laxhmi Chellappa, MD PhD [1,2]; Thomas Götz, MD[1]; Patrick Brunner, MD[1]; Vera Knoblauch, PhD[1]; Anna Wirz-Justice, PhD[1]; Christian Cajochen, PhD[1]

[1]Centre for Chronobiology, Psychiatric Hospitals of the University of Basel, Basel, Switzerland; [2]The CAPES Foundation/Ministry of Education of Brazil, Brasilia - DF, Brazil

Published in the Journal „SLEEP", in December 2011
(Sleep. 2011 Dec 1;34(12):1699-709)

ABSTRACT

Study Objectives: Major depressive disorder (MDD) is often associated with disturbances in circadian and/or sleep-wake dependent processes, which both regulate daytime energy and sleepiness levels.

Design: Analysis of continuous electroencephalographic (EEG) recordings during 40 h of extended wakefulness under constant routine conditions. Artifact-free EEG samples derived from 12 locations were subjected to spectral analysis. Additionally, half-hourly ratings of subjective tension and sleepiness levels and salivary melatonin measurements were collected.

Setting: Centre for Chronobiology, Psychiatric Hospitals of the University of Basel, Switzerland.

Participants: Eight young healthy women and 8 young untreated women with MDD.

Interventions: N/A.

Measurements and Results: MDD women exhibited higher frontal low-frequency (FLA) EEG activity (0.5-5.0 Hz) during extended wakefulness than controls, particularly during the night. Enhanced FLA was paralleled by higher levels of subjective sleepiness and tension. In MDD women, overall FLA levels correlated positively with depression scores. The timing of melatonin onset did not significantly differ between the two groups, but the nocturnal secretion of salivary melatonin was significantly attenuated in MDD women.

Conclusions: Our data imply that young women with MDD live on a higher homeostatic sleep pressure level, as indexed by enhanced FLA during wakefulness. Its positive correlation with depression scores indicates a possible functional relationship. High FLA could reflect a use-dependent phenomenon in depression (enhanced cognitive rumination or tension) and/or an attenuated circadian arousal signal.

Keywords: Major depressive disorder, women, frontal low-frequency EEG activity, subjective sleepiness, extended wakefulness, circadian rhythms, sleep-wake homeostat

INTRODUCTION

Major depressive disorder (MDD) is often associated with a dysregulation in circadian rhythmicity and/or sleep regulation. Abnormal circadian rhythms in many variables have been reported over the years, ranging from core body temperature, neurotransmitters and hormones to physiology of the sleep-wake cycle itself (Boivin 2000; Germain and Kupfer 2008). Characteristics of the circadian system (amplitude, phase and/or endogenous period) can be measured under very stringent laboratory conditions using markers such as core body temperature or melatonin. Although both delayed and advanced phase have been found in patients with MDD, several studies using highly controlled protocols such as the constant routine (Buysse, Monk et al. 1995; Wirz-Justice 1995) or forced desynchrony routines (Koorengevel, Beersma et al. 2002) (Koorengevel, Beersma et al. 2003) could not confirm circadian phase changes in MDD. However, reduced circadian amplitude seems to be more generally present (Souetre, Salvati et al. 1989; Wirz-Justice 1995; Armitage 2007; Germain and Kupfer 2008).

The process S-deficiency hypothesis postulates a deficiency in the homeostatic build-up of sleep pressure during wakefulness in MDD, leading to a shallower dissipation rate of process S during sleep (Borbely and Wirz-Justice 1982; Kupfer, Ulrich et al. 1984; Wirz-Justice 1995; Armitage, Hoffmann et al. 2000; Wirz-Justice 2006). Changes in homeostatic sleep regulation can be quantified by spectral EEG correlates in the low frequency range (1-7 Hz) during sleep and wakefulness (Torsvall and Akerstedt 1987; Aeschbach, Matthews et al. 1997; Aeschbach, Matthews et al. 1999; Borbely and Achermann 1999; Finelli, Baumann et al. 2000; Makeig, Jung et al. 2000; Cajochen, Wyatt et al. 2002). The intensity of low-frequency EEG activity at the beginning of sleep is proportional to the duration of prior wakefulness, and is considered to reflect the homeostatic aspect of sleep regulation (Borbely 1982; Daan, Beersma et al. 1984; Tobler and Borbely 1986). During sustained wakefulness, EEG activity in the 1-7 Hz range increases and can predict the subsequent homeostatic increase in slow-wave activity (SWA, EEG power density between 0.75-4.5 Hz) during sleep (Cajochen, Khalsa et al. 1999; Finelli, Baumann et al. 2000; Cajochen, Knoblauch et al. 2001), a phenomenon that is particularly pronounced in frontal brain regions (Cajochen, Khalsa et al. 1999; Cajochen, Knoblauch et al. 2001). This increased propensity in frontal low-frequency EEG activity (FLA) during sustained wakefulness suggests that frontal regions are more susceptible to sleep deprivation effects than other cortical regions (Horne 1993; Cajochen, Khalsa et al. 1999).

The process S deficiency hypothesis for MDD has rarely been tested in either sleep or waking EEG. An early study found lower delta waves during sleep in depressed patients (Kupfer, Ulrich et al. 1984), which was later documented only in males with MDD (Armitage, Hoffmann et al. 2000). In untreated middle-aged depressives, there was no difference from controls in SWA during sleep (Landolt and Gillin 2005).

Similarly, EEG studies during wakefulness in depression are contradictory and inconclusive (Pollock and Schneider 1991). It is surprising that this has not received more attention, since waking EEG-derived indices are a desirable biological measure in psychiatric disorders, given its practicability, low budget, and possibility of a greater number of recording sites (Pollock and Schneider 1991).

Thus, here was aimed at investigating sleep-homeostat and circadian-related differences in the EEG during extended wakefulness in MDD and healthy women under very stringently controlled laboratory conditions.

Our main hypotheses were as follows:

1. Women with MDD undergo a deregulation of sleep-wake homeostasis in comparison to healthy women, as indexed by an altered time course in EEG power density in the 1-5 Hz range during 40 h of extended wakefulness, particularly in frontal derivations, which are more susceptible to the effects of prolonged wakefulness.
2. Based on these alterations in sleep-wake homeostasis, which can affect subjective parameters, women with MDD experience higher subjective sleepiness and tension levels during 40 h of extended wakefulness.
3. Women with MDD show attenuated amplitude and/or circadian phase advance or delay in the rhythm of melatonin secretion.

METHODS

Study participants

All study participants were recruited via advertisements at different Swiss universities and on online job advertisement pages for students. A total of 900 candidates were enrolled as potential participants, and all completed a general questionnaire on their health status, medication and shift work, as well as a Beck Depression Inventory (BDI), Pittsburgh Sleep Quality Index (PSQI), and a Chronotype questionnaires. Out of these candidates, 80 young women were interviewed (SCID-I), and 25 volunteers were selected for study participation. Of these 25, 16 young women (mean age 24±4.8y [SD]) participated in the study. Most of the potential participants were excluded since they had evening chronotype or high PSQI (>8). All women were experiencing an episode of a MDD when undertaking the study protocol and fulfilled the diagnostic criteria of MDD according to the DSM-IV-TR. The main reason to include only women was based on the greater prevalence rate of MDD (without co-morbidity) in women then men.

Since all of our depressed women were rather young, they did not have a long history of depression. The depressed participants experienced either the first or the second onset episode; none of them had been given psychiatric (including psychotropic drugs) treatment before the study. The episode duration was ≥ 2 weeks, according to DSM-IV-TR criteria (prior mean duration in month 11.18±7.6 months). They had no atypical symptoms, did not experience severe sleep problems, as measured by the Pittsburgh Sleep Quality Index (PSQI ≤ 8; mean PSQI 6.5±1.6) (Buysse, Reynolds et al. 1989), and did not exhibit any comorbid psychiatric DSM-IV-TR-disorder. Each participant underwent a clinical interview, which was performed by the same clinical psychologist (ABP). This interview comprised the structured clinical interview for DSM-IV Axis I Diagnoses of existing symptoms (SCID-I; mean: 5.15±0.37 SD) (Wittchen, Wunderlich et al. 1996), the Hamilton-17 scale, the structured interview guide for the Hamilton depression rating scale with atypical depression supplement (SIGH-ADS; mean for HAMD-17: 12.29±2.49 SD) (Ramos-Brieva and Cordero-Villafafila 1988; Janet, Williams et al. 2003), the Montgomery-Åsberg Depression Scale (MADRS; mean: 16.71±2.13 SD) (Montgomery and Asberg 1979) and the Beck Depression Inventory (BDI; mean value 21.29±6.84 SD) (Beck, Ward et al. 1961). Half of the 16 MDD women were allocated to a high sleep pressure protocol (i.e. 40 h of extended wakefulness under constant routine conditions), while the other half participated in a low sleep pressure protocol (to be reported elsewhere).

The control sample comprised 8 healthy young women (age range: 20-31 years; mean age 25±3.3years, without any sleep problems [mean PSQI 2±1.63] (Knoblauch, Martens et al. 2003; Birchler-Pedross, Schroder et al. 2009). All study volunteers underwent a physical examination as well as an interview about sleep quality, life habits and health state. They were free of any medication intake or treatment (except oral contraceptives) for ≥ 2 months; they had no neurological or sleep disorders. Sleep efficiency did not differ in the 2 groups (P=0.54) measured by wrist actigraphy (mean value for MDD 88.67±4.3 SD; for healthy 92.41±4.0 SD). Volunteers were included only if their clinical sleep EEG scoring had no pathological findings (apnoea/hypopnoea-index [AHI] < 10/h; periodic leg movements [PLM] index <10/h). To exclude chronotype-specific differences in circadian phase preference we selected only moderate chronotypes (morning-evening-type [M/E] questionnaire rating between 14 and 21 points) (Torsvall and Akerstedt 1980). Thus, chronotype was not significantly different between the 2 groups (controls 15.6±3.6 vs. depressive 16.1±1.3), nor was the body mass index (BMI, 21.2±2.5 for the depressive and 20.9±1.4 for the healthy volunteers). All participants were nonsmokers and without any drug abuse, as verified by urinary toxicological analysis sensitive for amphetamines, benzodiazepines, opiates and tetrahydrocannabinol (Drug Screen Card Multi- 6®, von Minden GmbH, Moers, Germany). Participants were also required to abstain from excessive caffeine and alcohol consumption and heavy physical exercise. They indicated ≤ 3 cups of caffeinated beverages per day and ≤ 10 glasses of alcohol per week. Other exclusion criteria were: shift work within 3 months and transmeridian flights within 1 month prior to the study. All women (from both depressed and healthy controls) started the study on days 1–5 after menses onset in order to complete the entire study block within the follicular phase. Three women with MDD and five control women used oral contraceptives. Thus, our study group included 8 young women with MDD and 8 healthy controls. While this sample seems rather low, the study was been carried out under very controlled laboratory condition of a constant routine (CR). In addition, prior to the study, participants were required to adhere to a regular sleep-wake cycle as verified by actimetry and sleep logs and all spent an adaption night in the laboratory. Thus, this procedure significantly reduced variability in the output measures.

All procedures conformed to the Declaration of Helsinki. The local ethics committee approved the study protocol, screening questionnaires and consent form (Munch, Knoblauch et al. 2004; Munch, Knoblauch et al. 2007), and all study participants gave signed informed consent.

Study design

Each participant was instructed to maintain a regular sleep–wake cycle (bed- and wake-times within 30 min of self-selected target time), verified by wrist activity monitors (Cambridge Neurotechnology®, UK) and sleep logs for one week prior to the "in laboratory" part of the study. The entire study design entailed 2 protocols, one for high sleep pressure conditions and one for low sleep pressure conditions, with 8 controls and 8 depressive volunteers in each protocol. Participants were assigned randomly to either the low and high sleep pressure protocol. The treatment order ("sleep deprivation" vs. "nap protocol") was counterbalanced in order to avoid possible order effects. Here, we focus only on the high sleep pressure protocol, which comprised an 8-h full polysomnography night in the laboratory, followed by 3.5 days consecutive days in the laboratory. During day 1, participants adjusted to the experimental dim light condition (<8 lux). After a second 8-h sleep episode, all volunteers participated in a 40-h sleep deprivation protocol under controlled conditions (constant routine) (Cajochen, Khalsa et al. 1999; Cajochen, Knoblauch et al. 2001; Knoblauch, Munch et al. 2005) (Duffy and Dijk 2002), followed by a recovery night. The timing of the 8-h sleep episode was calculated with respect to the midpoint of each individual's habitual sleep episode, as assessed by actigraphy and sleep logs during the baseline week. All wake episodes were spent under semi-recumbent constant routine conditions (<8 lux) during wakefulness, with a minor shift to supine posture during scheduled sleep episodes (0 lux) (Cajochen, Knoblauch et al. 2001).

EEG Recording, Subjective Ratings, and Melatonin during Wakefulness

The Karolinska Drowsiness Test (KDT) (Akerstedt and Gillberg 1990; Gillberg, Kecklund et al. 1994) was performed every hour during scheduled wakefulness, starting 1 hour after habitual wake time. During the KDT, volunteers were instructed to relax, to keep their eyes open, and to avoid movement for 3 min, during which they had to fixate on a 5-cm dot attached to the wall at 1.5 m distance. These instructions were intended to maximize signal quality. Waking EEG activity was recorded continuously during the 40 h of extended wakefulness, using the Vitaport Ambulatory system (Vitaport-3 digital recorder TEMEC Instruments BV, Kerkrade, the Netherlands). Twelve EEG derivations (F3, F4, Fz, C3, C4, Cz, P3, P4, Pz, O1, O2, Oz referenced against linked mastoids), 2 electrooculograms, one submental electromyogram, and one electrocardiogram were recorded. All EEG signals were filtered at 30 Hz (fourth-order Bessel-type antialiasing low-pass filter, total 24 dB/Oct), and a time constant of 1.0 second was used prior to online digitization (range 610 μV, 12 bit AD converter, 0.15 μV/bit; storage sampling rate at 128 Hz). The raw signals were stored online on a Flash RAM Card (Viking, Rancho Santa Margarita, CA, USA) and downloaded offline to a PC hard drive. EEGs were subjected to spectral analysis using a fast Fourier transform (10% cosine 2-second window), which resulted in a 0.5-Hz resolution. The 3-min EEGs during the KDT were manually and visually scored for artefacts (eye blinks, body movements, and slow eye movements) offline. Approximately 80% of the waking EEG data was used after rejecting for epochs with artefacts. The absolute EEG power densities were then calculated for artefact-free 2-s epochs in the frequency range of 0.5 to 20 Hz. For data reduction, artefact free 2-s epochs were averaged over 20-s epochs (Cajochen, Knoblauch et al. 2001).

Subjective sleepiness was assessed every 30 min on the Karolinska Sleepiness Scale (KSS) (Gillberg, Kecklund et al. 1994). Subjective tension was assessed by a 100-mm bipolar VAS at 30-min intervals. The participants were asked to indicate how they felt "at the moment" by placing a vertical mark on the VAS ranging from 0 ("worst ever") to 100 mm ("best ever"). A similar VAS rating for mood was also made (Birchler-Pedross, Frey et al. 2010).

Salivary collections for hormonal assays were scheduled during wakefulness at the same 30-min intervals as for the subjective ratings. A direct double-antibody radioimmunoassay was used for the melatonin assay (validated by gas chromatography–mass spectroscopy with an analytical least detectable dose of 0.65 pm/ml; Bühlmann Laboratories, Schönenbuch, Switzerland (Weber, Schwander et al. 1997)). The functional least-detectable dose using the less than 20% coefficient of interassay variation criterion was < 0.65 pg/mL, and individual serum and saliva melatonin profiles showed excellent parallelism (r=0.977– 0.999; slopes =0.21– 0.63) (Weber, Schwander et al. 1997).

Statistics

For all analyses, the statistical packages SAS® (SAS Institute Inc., Cary, NC, USA; Version 6.12) and Statistica® (Stat-Soft Inc., 2000–2004, Statistica for Windows, Tulsa, OK, USA) were used. Repeated measure analyses of variance (rANOVAs) with the between factor "group" (depressive *vs.* control). We also considered the within factor "derivation" (EEG-channels: F3, Fz, F4, C3, Cz, C4, P3, Pz, P4, O1, Oz, O2) and the within factor "time-of-day" (11 time points; The 3.75-h interval came about 150 min of wakefulness followed by 75 min of the corresponding scheduled sleep (nap). This duration allows starting the recovery night at the same clock time [circadian phase] as the baseline night, since it replaced the last scheduled nap). These 3 factors ("group", "derivation", and "time-of-day") were performed for each 0.5-Hz frequency bin separately in the range of 1-20 Hz. Since we did not observe consistent left-right changes in MDD women *vs.* control women, frontal (F3, Fz, F4), central (C3, Cz, C4), parietal (P3, Pz, P4) and occipital derivations (O1, Oz, O2) were collapsed per subject into a single frontal derivation (average [F3, Fz, F4]), a single central (average [C3, Cz, C4]), a single parietal (average [P3, Pz, P4]) and a single occipital derivation (average [O1, Oz, O2]). Frequency bins yielding significance for the interaction "group x derivation" were collapsed into frequency bands, averaged per 3.75-h bin per study volunteer, and subjected to a rANOVA with the factors "group", "derivation" and "time-of-day". Similarly, the 30 min subjective ratings and melatonin values were collapsed into 3.75-h time bins resulting in 11 time points and subjected to rANOVAs with the factors mentioned above. All P-values derived from rANOVAs were based on Huynh-Feldt's (H-F) corrected degrees of freedom (significance level: P < 0.05). Alpha adjustment for multiple comparisons was applied according to Curran-Everett (Curran-Everett 2000). Pearson's correlation coefficients were computed to compare individual FLA levels with depressions scores derived from the MADRS and Hamilton-7-Item scale in MDD women.

RESULTS

EEG during Wakefulness

Absolute spectral EEG power density for each frequency bin, for each derivation and for each derivation averaged over eleven 3.75-h time intervals yielded a significant "group" effect for the frequency bins between 1 and 2.5 Hz, a significant "derivation" effect for a broader frequency range 1-13Hz, ($F_{3,42} \geq 3.3$) and 14.2-20 Hz ($F_{3,42} \geq 4.7$), and a significant interaction "group" x "derivation" effect between 0.5 and 4 Hz ($F_{3,42} \geq 2.9$) P < 0.05; Figure 1). Similarly, when considering averaged derivations (frontal, central, parietal, occipital), a significant "group" effect was elicited between 0.5 and 2 Hz ($F_{1,14} \geq 4.6$), and a significant (i.e. P < 0.05) "derivation" effect was elicited between 1-8 Hz ($F_{3,42} \geq 5.8$), 9.5-12 Hz ($F_{3,42} \geq 6.4$) and 16.5-20 Hz ($F_{3,42} \geq 4.0$).

Furthermore, the interaction term "group" x "derivation" yielded significant differences between 0.5 and 5 Hz ($F_{1,3} \geq 1.0$). Thus, EEG power density in the 0.5- 5 Hz range was collapsed per subject in order to investigate the time course of low-frequency EEG activity in the course of the 40-h episode of extended wakefulness.

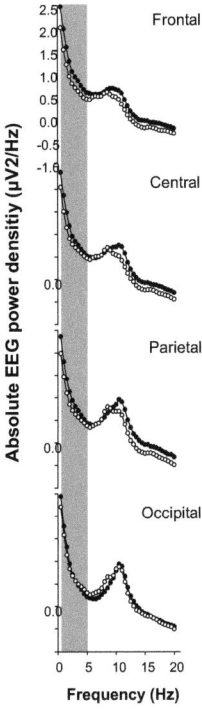

Figure 3.1: Absolute EEG power spectra during extended wakefulness along the anterior-posterior axis (frontal, central, parietal, occipital). Women with MDD are indicated by closes dots and the control group by open symbols. Mean values are shown for each 0.5-Hz frequency bin in the range from 0.5 to 20 Hz. A significant "group x derivation" effect was observed between 0.5 and 5 Hz, and a significant group effect in the range of 0.5-2 Hz.

Overall, low-frequency EEG activity (0.5- 5 Hz) showed a similar time course in both women with MDD and control women (Figure 2), with no significant differences in the interaction terms "group" x "time" and "group" x "time" x "derivation", although the interaction "group" x "derivation" yielded significant differences ($F_{3,42}= 2.9$; $P < 0.05$), indicating a frontal predominance of the increase in low-frequency EEG activity in MDD women compared to the control women (Figure 2).

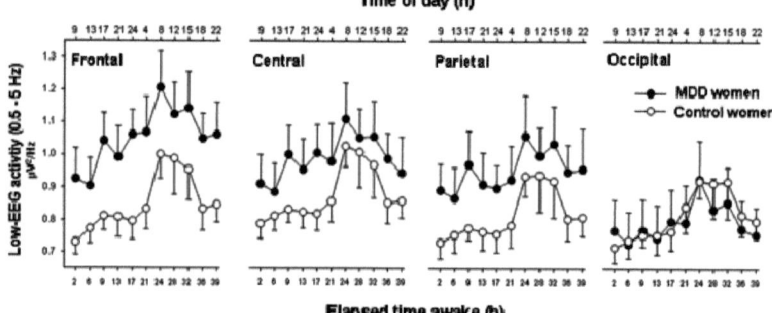

Figure 3.2: Dynamics of low-frequency EEG activity (EEG power density in the 0.5-5 Hz band) during 40 h of etended wakefulness. Data were binned into 3.75-h time intervals (mean values ±SEM, n=8) and plotted against relative time of day (h). Relative clock time represents the average clock time at which the time inetervals occurred. For statistics see text. Local derivations on waking-EEG are summarized as frontal derivation (mean of F3, F4, Fz) central derivation (mean of C3, C4, Cz), parietal derivation (mean of P3, P4, Pz), and occipital derivation (mean of O1, O2, Oz).

For enhanced visual illustration, a global cortical contour plot with the entire topography is provided in Figure 3 for EEG power density in the 0.5-5 Hz range. Average low-frequency EEG activity indicated higher values in frontal and central derivations in the MDD than control women.

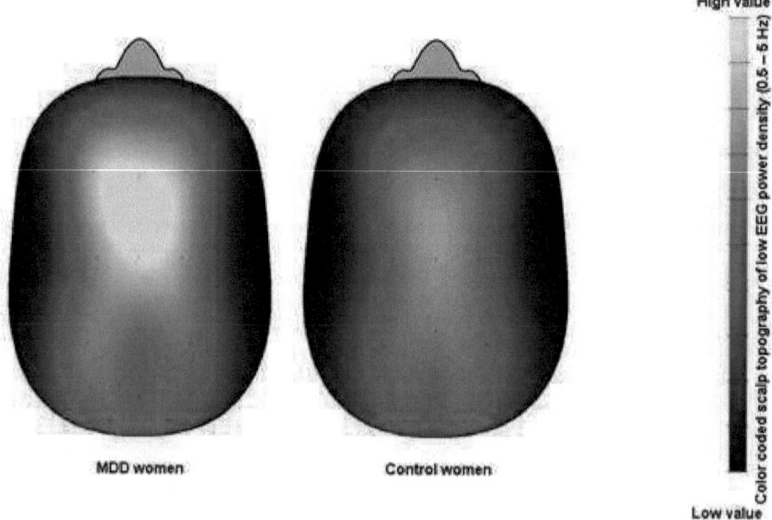

Figure 3.3: Colour-coded scalp topography of low-frequency EEG power density in the 0.5-5 Hz range averaged over 40 h of extended wakefulness in women with MDD (left head, n = 8) and control women (right head, n = 8). High temperature (yellow) represents high low-frequency EEG power density values, while low temperature (red) represents low values. Note: higher low-frequency EEG activity in MDD than control women, particularly in frontal and central brain regions.

Visual inspection of the contour plot over time (Figure 4) showed that MDD women had particularly high low-frequency EEG activity during the subjective night and early morning, as well as at 15:00h on the second day of extended wakefulness.

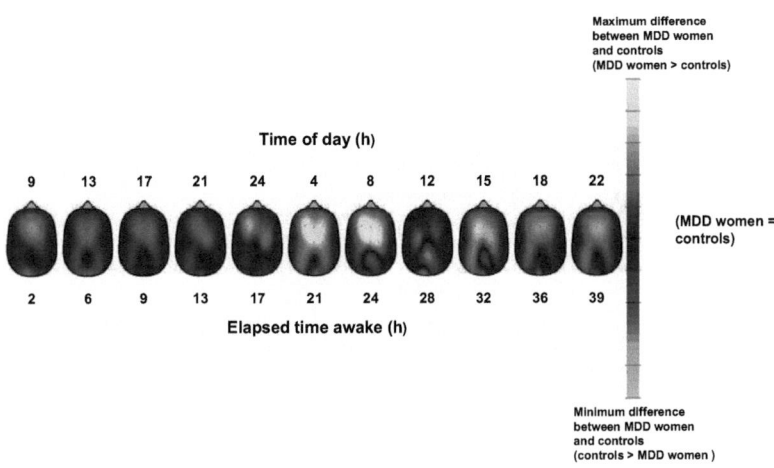

Figure 3.4: Time course of low-frequency EEG activity (in the range of 0.5-5 Hz) during 40 h of extended wakefulness. These contour plots (i.e. haeds) describe the difference between low-frequency EEG activity in the MDD and control women across 11 time points in the high sleep pressure protocol. MDD women showed higher low-frequency EEG activity, particularly during the subjective niaght and early morning, as well as at 15:00 during the second day. Maximum difference between MDD women and controls shown in yellow (max: MDD women > controls) and the minimum difference between the 2 groups is shown in light blue (max: controls > MDD women).

Subjective Sleepiness

The time course of subjective sleepiness ratings for MDD and control women during the 40-h of extended wakefulness are illustrated in Figure 5 (upper panel). The factor "group" yielded a tendency for higher sleepiness levels in MDD women ($F_{1,15}= 3.6$; $P = 0.07$), and significance for the factor "time-of-day" ($F_{10,150}=17.1$; $P<0.001$), the latter showing the expected circadian and wake-dependent modulation of sleepiness in both groups. Certain time points (from 17 h, 21 h and 24 h elapsed time into protocol) during the biological night yielded significant higher sleepiness levels in the MDD than in the control women ($F_{1,15}= 6.8$; $P < 0.02$).

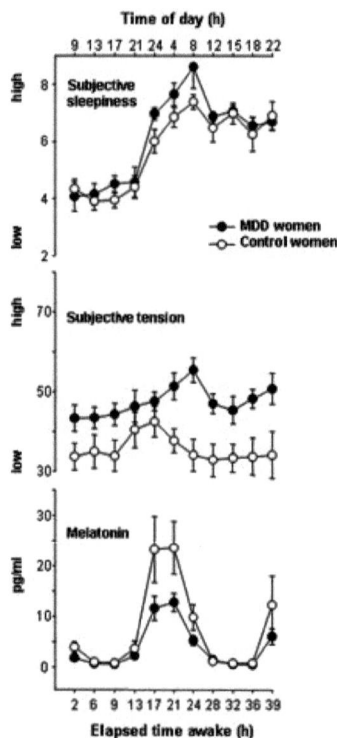

Figure 3.5: Time course of subjective sleepiness (Karolinska Sleepiness Scale, first panel), subjective tension (VAS, second panel), and salivary melatonin levels (third panel) across the 40 h of extended wakefulness. Women with MDD are plotted in closed dots, the healthy controls with open symbols (mean values ± SEM, n = 8).

Subjective Tension

MDD women indicated significantly higher levels of subjective tension only after 20 h to 40 h of wakefulness compared to the control women (Figure 5, middle panel). Thus, the factor "group" yielded significance ($F_{1,15}$= 12.9; P < 0.03), although no differences were observed for the factor "time-of-day" ($F_{10,150}$= 0.81; n.s.) and the interaction "group" x "time-of-day". However, certain time points from 17 h, 21 h, 24 h, and 28 h elapsed time into protocol) during the biological night and next morning were significantly higher in the MDD women than in the control women ($F_{1,15}$= 18.75; P = 0.0005).

Melatonin

The time course of salivary melatonin levels is illustrated in Figure 5 (lower panel). The factor "group" yielded a tendency for lower melatonin levels in MDD than control women ($F_{1,15}$= 3.7; P = 0.06). The factor "time-of-day" yielded significance ($F_{10,150}$= 16.2; P < 0.001), and the interaction term "group" x "time-of-day" almost reached significance ($F_{10,150}$= 1.9; P = 0.05), indicating an attenuation of nocturnal salivary melatonin secretion in the MDD women.

FLA and depression scores

Pearson correlations revealed positive and significant correlations between FLA levels during extended wakefulness and depression baseline severity, derived from both the Hamilton-7 Items (r = 0.75; P < 0.04) and MADRS (Figure 6; r = 0.8; P < 0.02).

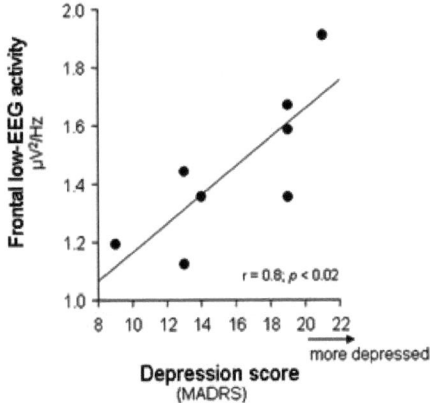

Figure 3.6: Linear regression between the values of the depression score (MADRS; 8-20) on the x-axis and the value of FLA (y-axis) across 40 h of extended wakefulness (r = 0.8; P < 0.02).

DISCUSSION

Overall FLA during extended wakefulness was higher in MDD than in healthy control women and correlated positively and significantly with depression severity. The time course of enhanced FLA was paralleled by higher subjective sleepiness and tension levels in MDD than in healthy control women. MDD women did not differ from controls in circadian melatonin phase, but showed a significant attenuation of melatonin secretion during the biological night.

These are the first data in MDD addressing the waking aspect of the S deficiency hypothesis postulated by Borbély and Wirz-Justice (1982) (Borbely and Wirz-Justice 1982). Surprisingly, unmedicated, young women with MDD, a middle chronotype and almost no sleep disturbances seem to live on a higher - not the hypothesized lower - homeostatic sleep pressure, but with similar build-up rates during extended wakefulness as found for healthy controls. The sleep aspect of the S-deficiency hypothesis was also studied in this same cohort (to be reported elsewhere (Frey, Birchler-Pedross et al. 2010)), and they indeed showed elevated – not diminished - SWA levels during sleep. To our knowledge, there is only one early study that reported elevated SWA during wakefulness in right frontal brain regions in a mixed-gender group of MDD patients with a comparatively brief medication washout period (Knott and Lapierre 1987). Together with the significant correlation with the MADRS and Hamilton depression scores, elevated FLA levels in our MDD group women most likely reflect certain aspects of the disorder *per se,* and do not seem to be primarily caused by a sleep disorder or a major circadian misalignment. Thus, we speculate that the elevated FLA levels during wakefulness in an episode of major depression were a state rather than a trait marker in our MDD cohort. Support for this assumption also comes from elevated SWA levels during sleep in the same MDD women (Frey, Birchler-Pedross et al. 2010). Our study sample comprised women with MDD who are unmedicated and without sleep disturbances. While this may not be a representative sample of patients with major depression, it should be emphasized that MDD itself is a heterogeneous group with symptoms that crucially depend on numerous aspects, such as duration of disorder. Most importantly, we could show that even in patients with MDD without medication and sleep disorders the homeostatic sleep regulation is significantly changed. In this context, one may speculate that changes in sleep homeostasis may anticipate and/or trigger a severe MDD episode, especially when considering that our sample included women with a mild episode of MDD. Similarly, FLA could also be seen as a compensatory mechanism and/or reaction to increased depressive levels, as indexed by an "over response" of the homeostatic sleep process when challenged by 40 h of sleep deprivation.

Interestingly, there is more recent works, running in a similar direction, showing that suppressing low-frequency activity in the EEG during sleep in MDD patients leads to mood improvements in those patients (Landsness, Goldstein et al. 2010). On the other hand, enhancing low-frequency activity during sleep marginally decreased positive mood in patients with MDD (Cheng, Goldschmied et al. 2010). These two studies lend support to our finding that the amount of FLA is significantly related to depression severity.

The topographic specificity of the increase in low-frequency activity – mainly frontal derivations – is an indication that women with MDD may be more susceptible to homeostatic sleep pressure, in particular the effects of prolonged wakefulness, possibly due to a high "recovery need" of frontal heteromodal association areas of the cortex, which are strongly affected by elevated sleep pressure as shown in PET studies (Thomas, Sing et al. 2000). Alternatively, our results of higher FLA could represent a biological correlate of higher cognitive rumination in MDD patients, known as "brooding", which is a core process in the onset and maintenance of depression (Nolen-Hoeksema 1991; Nolen-Hoeksema 2000) Depressive rumination could be seen as a state of higher arousal during wakefulness, which could have resulted in elevated subjective tension. If slow wave homeostasis is associated with net synaptic strength, postulated by Tononi et al. to increase during wakefulness and decrease during sleep (Tononi 2009), then higher cognitive rumination should lead to higher FLA during wakefulness, and FLA could be a proxy of more intense upscaling of synaptic strength in MDD. Thus, one could speculate that our women with MDD had an altered regulation of synaptic plasticity. Results from a recent study indeed support the hypothesis of a decreased synaptic plasticity in patients with MDD (Nissen, Holz et al. 2010).

Enhanced FLA levels impact on the circadian timing system. There is recent evidence for a crosstalk between systems regulating sleep-wake homeostasis and endogenous circadian rhythmicity in the hypothalamus (Schmidt, Collette et al. 2009). High sleep pressure levels considerably suppressed activity in the anterior hypothalamus, including the suprachiasmatic area, the brain site of the central circadian pacemaker (Schmidt, Collette et al. 2009). Thus, the observed decrease of circadian melatonin secretion in our MDD cohort could reflect an attenuation of the circadian wake-promoting signal by increased homeostatic sleep pressure. This finding is consistent with previous forced desynchrony studies in seasonal affective disorder (SAD) patients (Koorengevel, Beersma et al. 2002; Koorengevel, Beersma et al. 2003) and non-circadian protocols with depressed patients (Tononi 2009; Cheng, Goldschmied et al. 2010; Landsness, Goldstein et al. 2010).

It may be that low nocturnal melatonin levels and indeed low amplitude in many other variables (Wehr and Wirz-Justice 1982; Knott and Lapierre 1987; Strogatz, Kronauer et al. 1987; Dijk and Czeisler 1995; Lavie 2001; Koorengevel, Beersma et al. 2002; Koorengevel, Beersma et al. 2003) are a reflection of diminished circadian signal in MDD. Lack of an adequate wakefulness signal may permit the expression of increased FLA and sleepiness in these patients. On the other hand, one could also argue that, since our results on the circadian system focus only on melatonin as a circadian marker that the attenuated profile of melatonin in MDD could also reflect impaired nocturnal melatonin secretion. Taken together, these findings provide a possible explanation as to why bright light exposure, which increases the amplitude of the circadian system, may improve depressive symptoms in MDD (Shirani and St Louis 2009).

Limitations

Measuring behaviour under highly controlled laboratory conditions is pivotal to assess the contributions of circadian and homeostatic processes to the EEG during wakefulness. Our sample comprised young, unmedicated, depressed women, and excluded those with sleep disturbances; thus, our findings are not representative for all patients suffering MDD. Likewise, the directionality of the results (MDD prior to changes in sleep homeostasis and vice-versa) cannot be predicted by this study design, since a cross-sectional design and a constant routine protocol cannot provide a causal effect for these results. However, since the only difference between the depressive and control women was the depression *per se*, we could ideally test whether the observed changes are related to circadian and/or sleep homeostatic alterations.

CONCLUSION

The concomitant findings of higher FLA, subjective sleepiness, tension, and attenuated melatonin levels in untreated young women with MDD under stringent laboratory conditions indicate that major depression *per se* is associated with impaired night-time melatonin secretion and altered sleep-wake-homoeostatic processes.

ACKNOWLEDGMENTS

The authors thank Claudia Renz, Marie-France Dattler, Giovanni Balestrieri, and the student workers for their help in data acquisition. The authors also thank all the study volunteers for their participation in our study, Marcel Hofstetter for developing the software for the EEG topographical mapping, and to Dr. Antoine Viola for statistical advice.

This research was supported by Swiss National Science Foundation Grants START # 3100-055385.98, and 3130-0544991.98 and 320000-108108 to Dr. Cajochen, the Velux Foundation (Switzerland), the Daimler-Benz Foundation (Germany) and Bühlmann Laboratories, Allschwil (Switzerland).

CHAPTER 4

Diurnal mood variations in untreated unipolar depressed women: effects of sleep deprivation and multiple napping; under high and low sleep pressure conditions

Angelina Birchler Pedross, Sylvia Frey, Thomas Götz, Patrick Brunner, Vera Knoblauch, Anna Wirz-Justice and Christian Cajochen

Centre for Chronobiology, Psychiatric Hospital of the University of Basel, CH-4025 Basel, Switzerland

Manuscript prepared for submission (publication)

Abstract

Background: Diurnal mood variations are a key symptom in depression, but also occur in healthy people. Here, we aimed to investigate repercussions of differential sleep pressure levels [high (i.e. total sleep deprivation, SD) vs. low (i.e. multiple naps)] on diurnal mood levels in young women diagnosed with major depressive disorder (MDD) and healthy controls.

Methods: Eight healthy women (25±3.3y) and eight women (24±4.8y) with MDD underwent a 40-h SD (high sleep pressure) and a 40-h nap (low sleep pressure attained with multiple naps) protocol under constant routine conditions in the chronobiology laboratory. The volunteers rated their subjective mood every 30 minutes during scheduled wakefulness using a 100mm bipolar visual analogue scale. MDD women suffered from symptoms of "sadness", "loss of interest", "loss of energy", "reduced feeling of self-worth", "diminished concentration" and "social withdrawal" but not from insomnia as diagnosed by the Pittsburgh Sleep Quality Index and a full night sleep polysomnography.

Results: Under high sleep pressure conditions, diurnal mood variations were significantly modulated by the factors "group" (depressives *vs.* healthy), "time of day" (11 time points) and their interaction. Under low sleep pressure only a significant group effect was found. Overall, in both protocols, mood ratings were considerably lower in depressive women than in the control group. However, under SD women with MDD exhibited a more distinct circadian modulation of lower mood than controls. Despite this higher circadian variation of mood fluctuation, they did not profit from a significant and clinically relevant antidepressant effect of SD. Our data imply that depressive women without sleep disturbances show a significantly different time course of mood during elevated sleep pressure (i.e. 40-h SD) than controls but not during low sleep pressure conditions (i.e. 40-h nap condition). The lack of a clear antidepressant effect suggests that the SD response could depend on the magnitude of insomnia, which is a frequent co-morbidity in depression.

Key words: depression, sleep deprivation, constant routine, nap

INTRODUCTION

Total sleep deprivation (SD, "wake therapy") in depressed patients with severe insomnia revealed unexpected and paradoxical antidepressant effects (Reinink, Bouhuys et al. 1990). Thus, SD usually consisting of total sleep deprivation for one night or for the second half of a night, has been described as the most rapid antidepressant known, producing marked improvement within hours in approximately 60% of depressed patients (reviewed in (Wirz-Justice, Benedetti et al. 2005).. Cross-sectional studies showed that diurnal variation of mood (DV), a key symptom of depression, appears to be a predictive patient characteristic such that patients with marked positive DV (better in the evening) tend to respond more favourably to SD than those with negative DV (worse in the evening) (Haug 1992; Haug and Wirz-Justice 1993). Even though the improvement following total SD is rapid, there is also a rapid return of depressive symptoms after subsequent sleep recovery. The rapid, usually short-lasting improvement following total SD and the rapid return of depressive symptoms after subsequent sleep recovery indicates that the depression may be linked to sleep-wake regulating processes (Wirz-Justice and Van den Hoofdakker 1999). Additionally, SD needs to coincide with an early morning circadian phase for an optimal antidepressant response (Wirz-Justice 1998). Thus, depression may be closely linked to circadian sleep-wake regulatory processes. Based on this assumption, a chronobiological model was proposed to explain sleep-wake cycle disturbances in major depression on the basis of the two-process model of sleep-wake regulation (Borbely and Wirz-Justice 1982). This model attempts to explain the effects of SD, REM-sleep disinhibition and reduction of slow wave sleep (SWS) often seen in depressed patients as the result of a deficient sleep homeostatic process that leads to deteriorations in mood regulation (Borbely and Wirz-Justice 1982).

Indeed, sleep disturbances are one of the predominant symptoms associated with depression. Approximately 50-90% of depressed patients complain of difficulty falling asleep, staying asleep or experiencing early morning awakening or impairment of sleep quality (Casper, Redmond et al. 1985; Riemann, Berger et al. 2001). However, it is not clear whether the antidepressant effects of SD are linked to amelioration in the proposed S deficiency (Borbély and Wirz-Jusitce) and thereby short improvements in sleep quality in depression or whether SD acts an antidepressant independent on the patients sleep quality. To our knowledge, there are no studies that describe SD as an antidepressant in depressive patients without sleep disturbances. Here we investigated whether differential sleep pressure levels (high vs. low), as a marker of the sleep homeostatic process, have significant repercussions on self-rated mood levels in young women diagnosed with major depressive disorder (MDD) without sleep disturbances compared to age-matched healthy control women.

Our main hypotheses were as follows:

1. Depressive women exhibit an improvement in subjective mood during the course of a 40-h SD (high sleep pressure conditions) compared to control women.
2. Under low sleep pressure conditions (40-h nap protocol), subjective mood of depressive women undergoes a more pronounced circadian modulation than under high sleep pressure, but does not show improvement compared to controls.

METHODS
Study participants

All study participants were recruited via advertisement at different Swiss universities and on online job advertisement pages for students (for details see (Birchler-Pedross, Frey et al. 2011) . Sixteen young women (mean age 24±4.8y) participated in the study. All women fulfilled the complete diagnostic criteria of DSM-IV for MDD at the time of undertaking the study protocol. They had no atypical symptoms or other co-morbid psychiatric DSM-IV-disorder, and were without severe sleep problems as measured by the Pittsburgh Sleep Quality Index (PSQI≤8) (Buysse, Reynolds et al. 1989); mean PSQI 5.5±1.6SD. Each participant underwent a clinical interview with the same clinical psychologist (ABP). This interview comprised the structured clinical interview for DSM-IV Axis I Diagnoses of existing symptoms (SCID-I; mean: 5.2±0.4 SD) (Wittchen, Wunderlich et al. 1996). The study volunteers had all of the following symptoms at the time of the SKID Interview: "sadness", "diminished interest or pleasure", "energy loss", "reduced feeling of self-worth", "diminished concentration", and "social withdrawal"; six women described very moderate symptoms of "sleep disturbances". Clinical status was further estimated by the Structured Interview Guide for the Hamilton Depression Rating Scale with Atypical Depression Supplement (SIGH-ADS) (Janet, Williams et al. 2003), which consists only of the Hamilton-17 item scale (mean: 12.39±2.5 SD) plus atypical items (Standard Value for HAMD-17 ≥8), the Montgomery-Asberg Depression Scale (Montgomery and Asberg 1979) (MADRS: mean 16.7±2.1) (Standard Value for MADRS ≥ 13), the Beck Depression Inventory (Beck, Ward et al. 1961) (BDI: mean 21.3±6.8) (Standard Value for BDI ≥12). The "Mehrfachwahl-Wortschatz-Intelligenz-Test" (MWT-B: mean 30.4±2.5) was used for assessing intellectual ability. Two weeks after study completion, there was a follow up assessment of the BDI (mean 22.5±11).

The age-matched control group comprised eight healthy young women (age range 20-31y; mean 25±3.3y) who were medically screened and had no prior psychiatric illness (for further detailed information on the recruitment of the control women see (Knoblauch, Martens et al. 2003; Birchler-Pedross, Schroder et al. 2009)). None of the study volunteers were taking any medication, nor were they undergoing any kind of treatment. All study volunteers were free of neurological and other sleep disorders, as assessed by a full night polysomnographic screening (PSG). To exclude chronotype-specific differences in circadian phase preference, only moderate chronotypes (morning-evening-type questionnaire rating between 14 and 21 points (Torsvall and Akerstedt 1980) were selected. However, depressive study volunteers tended to be more evening types (16.1±1.3) than controls (15.6±3.7; p=0.066, t-test), although this difference was rather small. Body mass index (BMI) did not significantly differ between the groups (21.2±2.5 for depressive study volunteers and 20.9±1.4 for the healthy volunteers). All participants were non-smokers without any drug abuse, and were also required to abstain from heavy physical exercise, shift work and transmeridian flights before the study. All women started the study on days 1–5 after menses onset in order to complete the entire study block within the follicular phase, with the exception of three depressive and five control women taking oral contraceptives. All procedures conformed to the Declaration of Helsinki and the local Ethical Committee.

Protocol and study design

Each participant was instructed to maintain a regular sleep–wake cycle (bed- and wake-times within ±30 min of self-selected target time), which was verified by wrist activity monitors (Cambridge Neurotechnology®, UK) and sleep logs during one week prior to the study. The entire study design entailed two protocols; one for high and one for low sleep pressure conditions. In each protocol we included eight control and eight depressive volunteers. The eight healthy control women carried out a crossover design participating in both protocols. In the depressive cohort, we recruited 16 study volunteers and distributed them randomly to either the high or low sleep pressure protocol. Each protocol comprised 3.5 days and started with an 8-h PSG night in the laboratory. During day 1, the study volunteers adjusted to the experimental dim light condition (<8 lux). After a second 8-h sleep episode, the volunteers participated in either a high sleep pressure condition (40-h SD) or in a low sleep pressure condition (40-h multiple nap protocol) under strictly controlled constant routine (CR) conditions (Cajochen, Khalsa et al. 1999; Cajochen, Knoblauch et al. 2001), which was followed by a 8-h recovery night sleep. The timing of the 8-h sleep episode was calculated with respect to the midpoint of each individual's habitual sleep episode as assessed by actigraphy and sleep logs during the baseline week. All wake episodes were spent under semi-recumbent CR conditions (<8 lux) during wakefulness with a minor shift to supine posture during scheduled sleep episodes (0 lux).

Mood ratings

During the protocol, mood was assessed among the MDD women by means of different mood self- and foreign rating scales such as the Adjective Mood Scale (AMS, in its German version

"Befindlichkeitsskala") (Von Zerssen and Koeller 1976), the Montgomery Asberg Scale (Montgomery and Asberg 1979), the Hamilton-Depression Scale 7-Items German version analogous to (McIntyre, Konarski et al. 2005) and the Visual Analogue Scale (VAS) for subjective mood.

In this paper, we report results from the subjective mood (VAS) and the Hamilton-Depression scale 7 items (HAMD-7).

Subjective mood was assessed by a 100mm bipolar visual analogue scale (VAS) at 30-minintervals in both the control and MDD women. The participants were asked to indicate how they felt "at the moment" by placing a vertical mark on the VAS ranging from 0 ("worst ever") to 100mm ("best ever").

The observer's rating scale, [*Hamilton-Depression scale 7 items* (HAMD-7), the abbreviated HAMD-7 depression scale, which was essential because of its short duration in the elaborate constant routine protocol in the laboratory] was only assessed in the depressive women. The abbreviated HAMD-7 depression scale (McIntyre, Konarski et al. 2005) encompassed the items "depression mood", "feeling of guilt", "interest, pleasure, level of activity", "tension and nervousness", "physical symptoms of anxiety", "energy level", and "suicide". A score ≥ 4 indicates non/partial Remission, while a score of ≤ 3 indicates full remission. Important to notice is that this rating has no sleep item. This rating was carried out during the baseline week before the laboratory protocol started, and at 4 time points at about 2h (morning 9 a.m. ±1h), 14h (evening 8 p.m. ±1h), 26h (morning 9 a.m.±1h) and 36h (evening 8 p.m. ±1h) into the high or low sleep pressure protocol.

Data analyses and statistics

For all analyses, the statistical packages SAS® (Version 6.12; SAS Institute Inc., Cary, NC) and Statistica® (Stat-Soft Inc., 2000-2004, STATISTICA for Windows, Tulsa, OK) were used. For the observer and self-ratings, the SPSS® (SPSS Inc, SPSS for Windows, Chicago, IL, USA, and Version 17) was used. For data reduction, all values of subjective mood were collapsed into 3.75-h time bins per subject before averaging over subjects. Two-way measures ANOVA and the Procedure mixed with the factors "group" (depressive vs. control) and "time of day" (11 time points) were applied for each protocol (SD and NAP) separately. All p values were based on Huynh-Feldt's (H-F) corrected degrees of freedom (significance level $p<0.05$). At some time points data were not normally distributed, and thus a nonparametric test was used for *post hoc* comparisons (Mann-Whitney *U* test). The self- and observer ratings were analysed using a global mean for each protocol and then t-tests for independent samples at Levene's Test for Equality of Variance.

RESULTS

Subjective Mood levels under high sleep pressure conditions:
The time course of mood ratings for MDD and control women is illustrated in Figure 1. Under high sleep pressure conditions, variations in the time course of subjective mood was significantly determined by the main factors "group" (depressed vs. healthy women; p=0.0003), "time elapsed" (11 time points, p=0.0005) and the interaction of the two factors (p=0.04, Figure 1 left-hand panel). A post-hoc comparison yielded significantly worse mood in the MDD women at the time points 9h and 13h, as well as from time point 24h to 22h the next day.

Subjective Mood levels under low sleep pressure conditions: Under low sleep pressure conditions, subjective mood yielded a significant group effect (p=0.005), but neither a significant time effect nor a significant interaction effect (Figure 1, right-hand panel). In general, depressed women rated their mood significantly worse independent of time of day in the nap protocol, while both groups showed a distinct diurnal pattern with lowest mood ratings during the biological night, when endogenous melatonin values were highest (Frey et al. 2012).

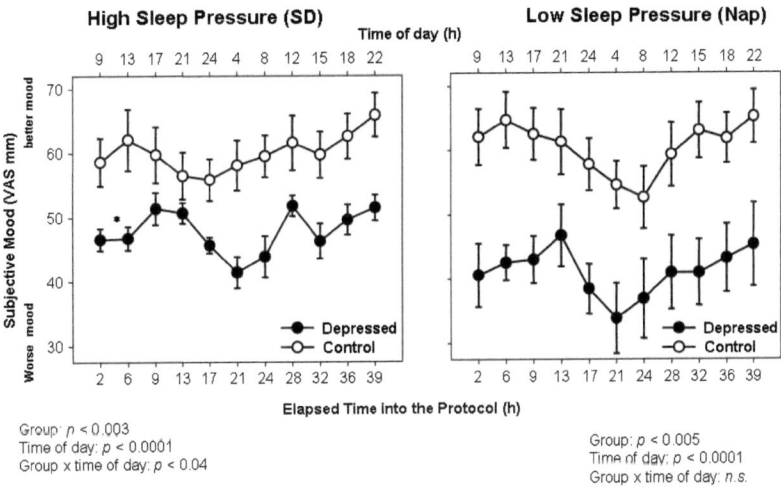

Figure 4.1: Time course of subjective mood (100mm visual analogue scale) during the 40-h high sleep pressure protocol (left-hand side panel) and during the 40-h low sleep pressure protocol (right-hand side panel). Mean values ± SEM (N=8 for controls and N=16 for MDD women). MDD women are plotted in

filled black symbols, the healthy control women with open symbols. The asterisks indicate the significant post-hoc values.

The Hamilton-7 Items observer rating yielded significance for the global mean between the high and low sleep pressure protocol (p=0.04, t-test), with higher values during the low sleep pressure protocol. A more detailed time course analysis, comparing each of the 5 time points between the two protocols revealed only one significant value at 22h; after 14 hours into the protocol (p=0.0315). No significant effect was found >20 hours of sustained wakefulness compared with baseline (Figure 2). Statistical analyses with non-parametric methods yielded similar results.

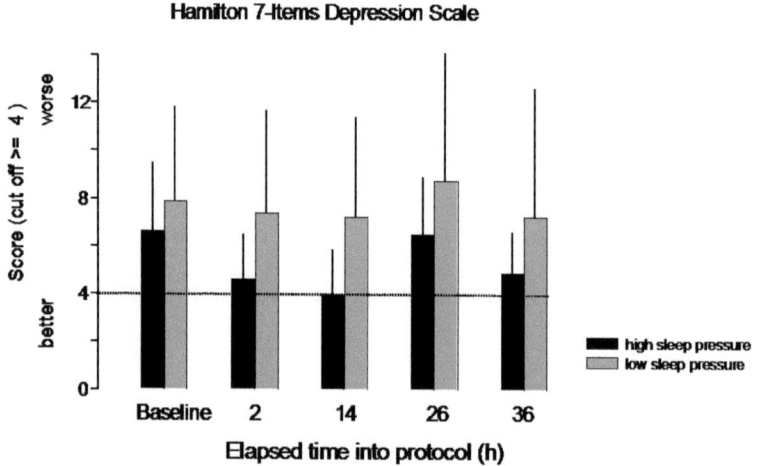

Figure 4.2: Estimation of depression levels by observer ratings using the Hamilton-7 Item short depression scale during a baseline (the night before the protocol start) and 4 time points during the high (black bar) and low (grey bar) sleep pressure protocol. Mean values + SEM.

A linear regression between the PSQI-values and subjective mood using the visual analogue scale (VAS) during high sleep pressure protocol at the time points "differences of point 6 and 8", i.e. after 21h minus 28h yielded a regression of r=0.75 (Figure 3).

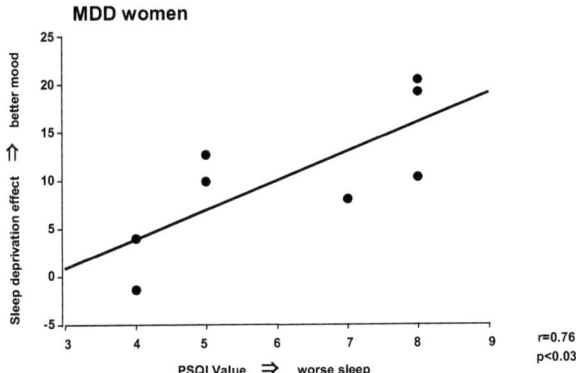

Figure 4.3: Linear Regression between the values of PSQI and the values of mood from the visual analogue scale (1 to 100) in the high sleep pressure protocol. This figure shows the difference between point 6 and 8, i.e. after 21h minus 28h during the 40h of wakefulness (see Figure 1 high sleep pressure protocol). On the x-axis PSQI values from 4 (the lowest value given) to 8 (the highest value given) are shown. On the y-axis, the difference between point 6 minus 8 of the mean values of the VAS of subjective mood are shown. Regression: $r^2=0.57$. The black dots are the individual PSQI minus mood point.

DISCUSSION

In this group of untreated young MDD women with a mild depressive episode, a middle chronotype and almost no sleep disturbances, total SD did not lead to a significant antidepressant effect when comparing subjective mood ratings of MDD women after >20 hours of sustained wakefulness with controls. This is the subjective mood rating of MDD women under high sleep pressure conditions compared to controls. Although there seems to be a very short-term improvement in mood observed at around 24h of elapsed time awake in the high sleep pressure protocol of subjective mood ratings when compared to time points in close vicinity. In the HAMD-7 Items rating, the time point after 14h of wakefulness was significantly better after SD when compared with sleep satiation. These discrete observable mood-elevating effects indicate that modulating sleep can have effects, albeit short, in depressed patients without sleep disturbances. However, the mood levels in the MDD women never attained the levels of the healthy control women. Likewise, the mood-elevating effects occurred only after 20 hours of wakefulness had elapsed, rather than continuously, as we would expect for a full SD effect.

These findings suggest that SD yields no antidepressant effect in young depressive women with major depression without insomnia. For this analysis we used two different kinds of measurement; a subjective visual analogue scale every half hour and one foreign rating every 12 hours. All measurements yielded a consistent result. The therapeutic effect of SD has often been described in terms of chronobiological models (Wirz-Justice, Tobler et al. 1981; Van den Hoofdakker, Beersma et al. 1986) and seems not to fit with our depressive cohort. According to different hypotheses of phase angle or circadian shift in non-seasonal depression and the reported results of SAD patients, (Graw, Haug et al. 1998; Koorengevel, Beersma et al. 2003) in our cohort we did not observe a significant difference between the MDD women and control volunteers (Frey, unpublished data). Our data questions the causes behind this non-significant antidepressive SD effect.

The fast but short-lasting improvement of depressive symptoms by SD in about 60% of patients with a major depressive disorder is well established, but the mechanisms of this are still not clear. A high delta sleep ratio is a positive predictor for SD response (Nissen, Feige et al. 2001), which could not be confirmed in our MDD women. Certainly, we found that the MDD women may live on a higher level of homeostatic sleep pressure manifested by EEG slow-wave activity (SWA 0.75-4.5 Hz) during baseline sleep, and recovery sleep from high and also low sleep pressure was significantly higher in the MDD women than in controls (Frey, Birchler-Pedross et al. 2010). Moreover, MDD women responded with a stronger EEG synchronization in a frequency range of 0.5-5 Hz during sustained wakefulness during the SD (Birchler-Pedross, Frey et al. 2011), which confirmed the higher delta waves during sleep, but not the role of delta sleep as a positive predictor for the SD response. In contrast, SWS changes in depression, was reported as reduced and SD for one night exerts an immediate antidepressant effect that is short lived. Thus, it was hypothesized that sleep regulation (Process S) is a deficient depression alleviator (Borbely and Wirz-Justice 1982).

The antidepressant effect of SD was attributed to the increased level of Process S attained by prolonging wakefulness. We could not confirm this hypothesis in our study, since the only grounds for this that we could determine could be due to the fact that our cohort of MDD women has no sleep disturbances. In this case, we can confirm part of the chronobiological model insofar as it is based on the theory of impaired sleep.

How could our findings be of every day value to the clinician ? In studies of physiological differences between insomnia with depression versus depression, we can conclude, based on these preliminary observations that insomnia with depression and depression alone appear to be two distinct entities. There is clinical and epidemiological evidence that sleep disturbances in depression constitute a risk factor for poor clinical outcomes. Specifically, insomnia complaints tend to precede the onset and recurrence of depression (Perlis, Giles et al. 1997; Cole and Dendukuri 2003; Riemann and Voderholzer 2003) in as many as 40% of cases (Ohayon and Roth 2003). The risk of developing major depression is significantly increased in individuals complaining of insomnia (Dryman and Eaton 1991; Breslau, Roth et al. 1996; Mallon, Broman et al. 2000). Sleep disturbances in depression can predict treatment outcomes. Specifically, poor sleep quality predicts poor response to non-pharmalogical treatments (interpersonal psychotherapy) of depression (Buysse, Kupfer et al. 1999). Finally, subjectively reported better sleep quality post-treatment is associated with lower rates of recurrence of depression (Buysse, Frank et al. 1997). Together, these observations suggest a critical role played by circadian and sleep disturbances in the pathophysiology of depression. Insomnia is thereby not only experienced subjectively but also reflected in altered objective sleep architecture, as first demonstrated by Kupfer and colleagues in the early 1970s (Kupfer and Foster 1972).

CONCLUSION

The constant routine study conditions allowed us to very precisely measure circadian rhythms in mood during a depressive episode. The women with major depression exhibited strong diurnal mood variations not found in healthy controls. Even though it has been shown that high variability of mood fluctuations is a good predictor of SD response, this group of MDD women without sleep disturbances did not respond to SD. These findings indicate that the antidepressant response to SD may depend on the magnitude of insomnia present.

ACKNOWLEDGEMENTS

We thank all our technicians Marie-France Dattler, Claudia Renz, Giovanni Balestrieri and the student shift workers for their help in data acquisition as well as to the study participants, and Sarah Chellappa for the fruitful discussions and comments on the manuscript.

This research was supported by Swiss National Science Foundation Grants START # 3100-055385.98, and 3130-0544991.98 and 320000-108108 to CC, the Velux Foundation (Switzerland), the Daimler-Benz Foundation (Germany) and Bühlmann Laboratories, Allschwil (Switzerland).

CHAPTER 5

Young depressed women perform faster in a psychomotor vigilance task during sleep deprivation than controls

Angelina Birchler Pedross, Sylvia Frey, Thomas Götz, Patrick Brunner, Vera Knoblauch, Anna Wirz-Justice and Christian Cajochen

Centre for Chronobiology, Psychiatric Hospital of the University of Basel, CH-4025 Basel, Switzerland

ABSTRACT

Background The psychomotor vigilance task (PVT) is a neurobehavioral task that can sensitively measure the effects of sleep pressure and circadian phase in healthy individuals. However, so far no studies have focused on PVT performance in depressive volunteers under differential sleep pressure conditions (sleep deprivation (SD) vs. sleep satiation (NAP). Here we examined the effects of high (40-h SD) and low sleep pressure (40-NAP) on PVT performance in young women with major depression in comparison to healthy controls.

Methods Eight healthy women (mean age 25±3.3y) and eight women (mean age 24±4.8y) with MDD underwent 40-h SD (high sleep pressure) and a 40-h multiple nap paradigm (low sleep pressure with intermittent 150/75 minutes of scheduled sleep-wake episodes) under constant routine conditions. In both protocols, PVT performance was regularly measured during 5 minutes in 11 tests conducted every 225 minutes.

Results Depressed women had significantly faster reaction time (RT), as indicated by significantly lower median, 10%-fastest and 10%-slowest RT (main factor 'group', p=0.05) during high sleep pressure conditions. No significant differences between the depressed and control women were elicited during low sleep pressure conditions in the nap protocol. The interaction "time-of-day" x "group" yielded no differences in both protocols, however, when considering time-of-day effects; depressed women were comparatively faster in the three abovementioned RT variables under both protocols.

Conclusion Depressed women had faster RT in PVT performance during sleep deprivation. This may imply that depressed women are less susceptible to the wake-dependent detrimental aspects on PVT performance than non-depressed women. In other words, PVT performance in depression may not be directly linked to well-known "prefrontal tiredness", but rather involve subcortical structures and/or reflect higher cognitive arousal caused by rumination.

Keywords: PVT, Major Depression Disorder, Sleep Deprivation, Constant routine. Nap

INTRODUCTION

Sustained attention is fundamental for optimal cognitive functioning. Neurobehavioral and cognitive functioning appear to be strongly influenced by two non-additive and interlinked systems: the circadian timing system and the homeostatic sleep regulatory system. The interplay of these two processes determines the timing, duration, and quality of sleep and wakefulness and leads to a characteristic circadian performance pattern in several neurobehavioral tasks, such as the psychomotor vigilance task (PVT). Circadian rhythmicity has been elicited for a wide range of neurobehavioral variables, such as vigilance and choice reaction time for a review see (Rogers, Dorrian et al. 2003), with a minimum close to the nadir of core body temperature.

Sleep deprivation can negatively impact on cognitive performance, particularly in cognitive tasks which rely on prefrontal cortex (PFC) function for review (Rogers, Dorrian et al. 2003). Continuous load on PFC activation enhances its susceptibility to „deterioration" with elapsed time awake (Harrison, Horne et al. 2000) which in turn can be associated with decreased higher cognitive performance, such as executive function (Drummond, Bischoff-Grethe et al. 2005), and also cognitive processes involving speed and attentional accuracy. Many previous studies have shown impairments on a diverse range of cognitive functions such as memory, planning and executive attention (Weingartner, Cohen et al. 1981). These cognitive functions belong to the brain region of frontal cortex. Prominent models of neurobiology of depression implicate involvement of the anterior cingulated cortex (ACC) and the dorsolateral prefrontal cortex (DLPFC).

Reaction time in PVT performance shows clear dose-response sleep deprivation effects, as indexed by lengthier RT with partial sleep deprivation, which becomes progressively more lengthened with longer deprivation periods (Dinges, Pack et al. 1997; Beaumont, Batejat et al. 2001). The theoretical underpinnings may involve decreased arousal as indicated by increased sleep pressure leading to inconsistent performance (greater response variability with lapses on RT and sustained attention tests) and wake-state instability. In the former, increased propensity to sleep may negatively impact on frontal lobes leading to behavioural lapses; response slowing, time-on-task decrements, and errors of commission (Saper, Chou et al. 2001). In the latter, cognitive performance variability may reflect wake state instability (Saper, Chou et al. 2001) according to which sleep-deprived individuals undergo reciprocally inhibiting neurobiological systems that mediate sleep initiation and wake maintenance (Mignot, Taheri et al. 2002). This can be indicated by vigilance tasks with longer RT and increased errors of omission at longer time intervals.

Taken together, the above mentioned data suggest that cognitive performance decrements in vigilance tasks depend heavily on the level of the sleep–wake homeostat. While there is substantial body of evidence in favour of performance decrements in healthy volunteers, it is striking that currently no studies have focused on the PVT performance in depressive volunteers under sleep deprivation (SD) condition. Thus, our aim was to elucidate how the sleep-wake homeostat and the circadian system can impact on PVT performance and subjective alertness in depressed individuals, under stringent controlled laboratory conditions, during 40-h of sleep deprivation and 40-h of sleep satiation (multiple nap protocol, NAP). We hypothesised as follows:

1. Depressed women show worse psychomotor vigilance performance than controls. However, this difference disappears with accumulating sleep pressure in the high sleep pressure protocol, due to the higher "prefrontal tiredness".
2. Psychomotor vigilance performance under low sleep pressure will be constantly reduced in depressed women compared to controls.
3. The improvement in psychomotor vigilance performance during high sleep pressure conditions in depressed women correlates with the antidepressant effect of sleep deprivation.
4. Depressed women will undergo inferior physical comfort than healthy controls during the PVT performance under high sleep pressure condition.

METHODS

Study participants

All study participants were recruited via advertisements at different Swiss universities and on online jobs advertisements pages for students. A total of 16 young women (mean age 26±5y) have participated in the study. All 16 young women were suffering from an episode of Major Depression Disorder (MDD) during the study protocol: they fulfilled the diagnostic criteria of DSM-IV for MDD, with no atypical symptoms, and without other co-morbid psychiatric DSM-IV-disorders. Each of the participants underwent a clinical interview including the SKID-Interview (Wittchen, Wunderlich et al. 1997), a structured interview for the Hamilton Depression Scale (SIGH-ADS, Hamilton-17) (Janet, Williams et al. 2003), Montgomery Åsberg Depression Scale (MADRS) (Montgomery and Asberg 1979), Beck Depression Inventory (BDI) (Beck, Ward et al. 1961). All the depressive subjects had a mean value of BMI 22.07±3.2SD and a mean of chronotype of 15.88±1.35SD.

The healthy cohort comprised eight young subjects (8 women age range 20-31 years; mean age 25±3.3years [SD]. For detailed information of the control subjects please see (Knoblauch, Martens et al. 2003).

All subjects non-smokers, free from medical, psychiatric, neurologic and sleep disorders (Pittsburgh Sleep Quality Index score<8) (Buysse, Reynolds et al. 1989)and average chronotypes (score between 14 and 21) as assessed by screening questionnaires, a physical examination and a polysomnographically recorded screening night. Inclusion criteria included volunteers with no sleep disorders[apnoea/hypopnoea-index (AHI) <10/h; periodic leg movements (PLM) index <10/h]. Participants were also required to abstain from excessive caffeine and alcohol consumption as well as heavy physical exercise. Other exclusion criteria were: shift work within 3 months and transmeridian flights within 1 month prior to the study. Subjects started the study on days 1–5 after menses onset in order to complete the entire study block within the follicular phase. Three depressive and five control subjects were taking oral contraceptives.

All procedures conformed to the Declaration of Helsinki. The local Ethical Committee approved the study protocol, screening questionnaires and consent form [for details see (Munch, Knoblauch et al. 2004)], and all study participants gave signed informed consent.

Protocol and study design

Each participant was instructed to maintain a regular sleep–wake cycle (bed- and wake-times within ±30 min of self-selected target time), which was verified by wrist activity monitors (Cambridge Neurotechnology®, UK) and sleep logs during one week prior to the study. The study design entailed two protocols: one for high and one for low sleep pressure condition. In each protocol, we included eight control and eight depressive subjects. During day 1 subjects adjusted to the experimental dim light condition (<8 lux); one depressive subject received a low-dose heparin injection on two consecutive days of each study block (Fragmin® 0.2 ml, 2500 IE/Ul, Pharmacia AG, Dübendorf, Switzerland) in order to prevent venous thrombosis. After a second 8-h sleep episode, all subjects participated in a 40-h sleep deprivation protocol under chronobiological controlled conditions as detailed in (Cajochen, Khalsa et al. 1999) which was followed by a recovery night. The 8-h sleep episode was calculated with respect to the midpoint of each individual's habitual sleep episode as assessed by actigraphy and sleep logs during the baseline week. All wake episodes were spent under semi-recumbent CR conditions (<8 lux) during wakefulness with a minor shift to supine posture during scheduled sleep episodes (0 lux).

Psychomotor vigilance task

The PVT task is a sustained attention performance task (Dinges, Pack et al. 1997) sensitive to sleep loss and circadian rhythmicity (Dinges, Pack et al. 1997). The study participants are required to quickly press a button as soon as red digits appear on a millisecond counter. The digits are presented in intervals randomly varying from 3 to 7 s. During both protocols, 11 test bouts of 5 minutes duration were performed every 225 min starting 75 min lights on in the morning. In the NAP condition the PVT was scheduled 75 min before and after each nap. All participants were instructed to press the response button as fast as possible as the red digits appeared.

Subjective alertness comprised subjective experience of feeling alert as indexed by the visual analogue scale "very alert to not at all alert", which can be assessed using likert-type self-rating scale such as the Karolinska Sleepiness Scale (KSS) (Akerstedt and Gillberg 1990).

Data Analyses and Statistics

We analyzed median RT, mean of 10% fastest, and of the 10% slowest RT's, all converted to reciprocal RT's. In addition, we calculated the interpercentile range between the 10th and 90th percentile according to (Graw, Krauchi et al. 2004). In a first step, a two way ANOVA analysed by Procedure Mixed with the repeated factors "elapsed time" (session 1 to 11) and the between factors "group" was performed for each protocol separately. All derived p-values were based on the Huynh-Feldt's corrected degrees of freedom Alpha criterion was set at p=0.05. SAS® (SAS Institue Inc., Cary, NC, USA; Version 9) and Sigma plot® to visualize data (SIGMA plot Version 9).

RESULTS

Psychomotor vigilance task (PVT): Main factor "group" yielded significant differences for global mean value of Reaction Time (RT) for both protocols (p=0.0097 for SD, p<0.001 for NAP). The Interpercentile range did not differ in either in both protocols, however, main factor group yielded significant differences for three dependent RT variables (median, 10%-fastest and 10%-slowest RT, p=0.05), with significantly faster reaction time for the depressive cohort in SD protocol, although no group-effect differences were elicited for the nap condition. Time of day yielded significant differences for both protocols, although the interaction of the "time of day" and "group" indicated no differences under both protocols.

Table 1 illustrates the results of the rANOVAs of the measures 'median RT', '10% slowest RTs', '10% fastest RTs', and the interpercentile range (10th–90th percentile).

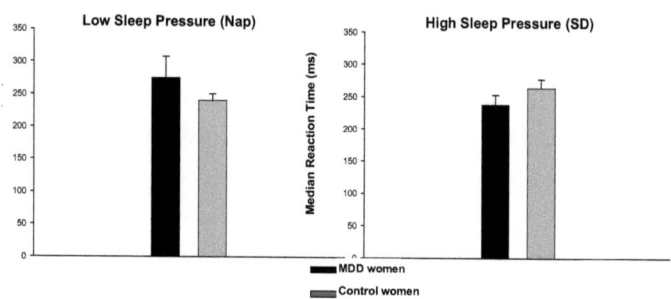

Figure 5.1: Median RT (ms) of the depressed and control group under low (Nap) and high (SD) sleep pressure conditions.

Factors	Reaction Time (Median, p50)	Reaction Time (10% fastest, p10)	Reaction Time (10% slowest, p90)	Reaction Time (Inter-percentile range 10th-90th)
Time of day (SD)	sign. p=0.0005	sign. p=0.0005	sign. p=0.0172	n.s.
Time of day (NAP)	sign. p=0.0002	sign. p=0.0006	sign. p=0.0028	n.s.
group (SD)	sign. p=0.048	sign. p=0.0266	sign. p=0.0417	n.s.
group (NAP)	n.s.	n.s.	n.s.	n.s.
Time of day x group (SD)	n.s.	n.s.	n.s.	n.s.
Time of day x group (NAP)	n.s.	n.s.	n.s.	n.s.

Table 5.1: Results of the r ANOVAs of the measures "median RT", 10%slowest RTs", "10%fastestRTs", and the interpercentile range (10th-90th percentile).

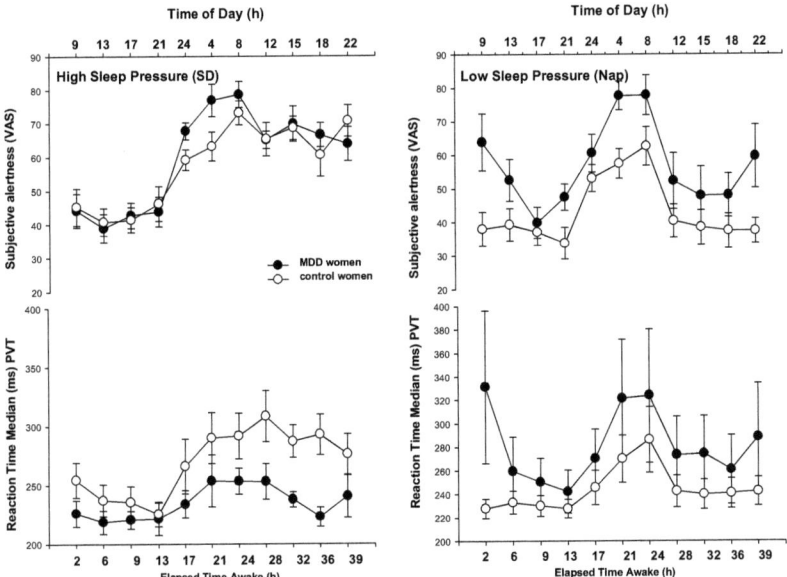

Figure 5.2: Time course of vigilance performance exemplified by median reaction time of depressed and the controls under low (NAP) and high (SD) sleep pressure conditions upper panel. Time course of subjective alertness (100- mm visual analogue scale) during the 40-h high sleep pressure protocol (upper panel) and during the 40-h low sleep pressure protocol (lower panel). Mean values ± SEM. Depressed patients are plotted in dots, the healthy controls with open symbols.

Subjective alertness (VAS): Regarding the time course of subjective alertness, the depressive young women rated their alertness not different from the control group during the high sleep pressure illustrated in the top panel of figure 2 neither and during the low sleep pressure illustrated in Figure 3 top panel, which devoted no statistically significant 'group' effect (p=0.5). The 'time of day' in subjective sleepiness yielded a very significant value (p<.0001), which shows the expected modulation of alertness during wakefulness in both groups. The interaction of the two-way rANOVA for the factors group and time of day was not significant (p=0.4).

DISCUSSION

The main finding of this study is that depressed women show a comparatively faster performance in the PVT under high sleep pressure condition. This is strikingly unusual when one has in mind that depression is more frequently associated with cognitive decrements, particularly in tasks involving PRC function. Furthermore, circadian rhythms in depression, as indicated by attenuated melatonin amplitude during sustained wakefulness (Birchler-Pedross, Frey et al. 2011), appear to indicate that depressed women may undergo a weakened circadian arousal signal that does not appropriately oppose the sleep homeostatic process, which is related to higher subjective tension. Considering these factors, why does PVT performance in our depressed subjects go in an opposite direction? Within a psychological framework, depressed patients with alertness problems frequently complain about higher levels of tiredness and lower capability to cope with stress, which can suggest that individuals who feel subjectively sleepier are also more cognitively impaired (Leproult, Colecchia et al. 2003). However, this assumption may not hold true. Accumulating evidence indicates that subjective sleepiness and objective alertness are not necessarily related to performance measures during sleep deprivation protocols (Frey, Badia et al. 2004) showed that depressive subjects need more time for cognitive decision but not for motor speed reaction time, which may partially underlie our findings. In other words, while depressed subjects experience symptoms of "loss of interest", "loss of energy", "loss of motivation" this does not necessarily imply that tasks focused on quick motor reaction times, such as very short RT of the PVT, will exhibit a decrement. We do not have the true explanation for the faster PVT performance in depressed young women. The faster performance may be correlated to the very short task duration of 5 minutes. It has to be assumed that depressive patient will be slower while the task duration expanses.

Other line of reasoning builds up from discrepancies between "sleepiness" and "alertness". These constructs are often assumed to be reciprocal states of consciousness e.g. (Moller, Devins et al. 2006). Alertness and sleepiness are not necessarily reciprocal states, since studies have indicated that subjective states of impaired alertness and excessive sleepiness are independent constructs in the evaluation of sleep-disordered patients e.g. (Moller, Devins et al. 2006), Sleepiness and alertness states are in the background of most cognitive processes, even if not systematically modulating performance. Measures of subjective sleepiness are generally sensitive to decreased alertness and performance decrements during sleep deprivation but are weakly correlated to performance on RT, vigilance and cognitive tasks in individuals (Binks, Waters et al. 1999). The effects of sleep deprivation on other cognitive tasks were less consistent than the effects on RT and vigilance tasks. To some extent this lack of consensus is due to different experimental conditions, length of testing battery, difficulty of the task. Drummond et al., (Drummond, Meloy et al. 2005) found that particularly poor performance after sleep deprivation may elicit a subsequent attentional recovery that manifests itself as greater activation within the brain regions normally responsible for fast reaction times. Furthermore, even factors such as body posture influence RT. RT performance in overnight sleep deprived subjects is slowed when seated, but normal when standing (Caldwell and Gilreath 2002). The effects of sleep deprivation are complicated by the interaction of diurnal variation (circadian phase) with monotonic effects (length of sleep deprivation), consistent with the two-process model proposed by Borbély (Achermann and Borbely 2003) (Cajochen, Khalsa et al. 1999). Also, effects of sleep deprivation on performance are not uniform and there is considerable inter subject variability, likely representing individual characteristics (Bonnet and Arand 2005).

A cortical hallmark of neurobehavioral functioning is the prefrontal cortex (PFC). A number of performance tasks thought to be putatively subserved by the PFC have been reported to demonstrate significant impairment during sleep loss, which is reversible following recovery sleep (Harrison, Horne et al. 2000). This has been corroborated by anatomical evidence that the central pacemaker located in the SCN plays a major role in the regulation of arousal and attention through noradrenergic mechanisms (Aston-Jones, Rajkowski et al. 1999) (Aston-Jones, Chen et al. 2001) (Doran, Van Dongen et al. 2001) (Harrison and Horne 1998). The SCN projects to the PFC (Sylvester, Krout et al. 2002), via its relay in the paraventricular thalamic nucleus (PVT) to the medical PFC in rats. Assuming that a homologous circuit exists in humans, these pathways may modulate high level brain functions, such as attention or working memory. This could explain the effects of circadian phase *per se* on neurobehavioral function and sleepiness, which have been reported in non-sleep deprived subjects during forced desynchrony protocols (Wyatt, Ritz-De Cecco et al. 1999) and multiple nap studies (Cajochen, Knoblauch et al. 2001) (Cajochen, Knoblauch et al. 2001); (Lavie 1986).

A further aspect to explain our conflicting results could be motivation during neurobehavioral performance. One psychological construct with is underlying brain systems impacting sustained attention is motivation. Based on animal as well as human lesion studies, the motivational system includes portions of the frontal lobes (e.g. anterior cingulated) as well as limbic and subcortical structures (striatum, nucleus accumbens, and amygdala) and much of the dopamine system (Robbins and Everitt 1999). Subjects who are apathetic or unmotivated will not be as vigilant as those with high motivation. Performing task with a high financial reward for performance engages the attention system stronger and with longer duration than performing the same task with no overt reward for performance (Homberg, Grunewald et al. 1981) (Begleiter, Porjesz et al. 1983). Conceptually, effort (Kahneman 1973) and motivation are related. Several factors modify the effects of sleep deprivation on RT and vigilance task performance. Long, simple low-interest, self-paced RT tests are most affected by sleep loss. The effects on RT are partially reversed when there is feedback provided when the task is shorter or more challenging, or when there is a reward used as incentive for performance, suggesting that increased motivation or interest can at least partially compensate for the effects of sleep deprivation (Steyvers and Gaillard 1993). Taken together, one might speculate that tasks of short duration may engage depressed subjects to obtain a better performance. Another explanation as to why depressive subjects are faster in PVT could be explained by the effects of Total Sleep Deprivation (TSD). TSD appears to be simultaneously arousing (energy-giving) and de-arousing (leading to less tension), while this response takes place against a background of increased tiredness/sleepiness. It is argued that TSD triggers a psychological disinhibition process on cortical fatigue, most possibly due to a dampening of subcortical arousal systems (Van Den Burg, Beersma et al. 1992) although prefrontal cortical areas may also be implicated. The rapid, usually-short lasting improvement of depressive symptoms (Wirz-Justice and Van den Hoofdakker 1999), which in turn could be an underlying reason for subjects to experience task engagement; can not explain this faster performance of the depressive subjects.

While speed performance on simple repetitive (Colquhoun 1981) and serial search tasks (Monk 1982) peaks with temperature levels in the evening, speed performance on more complex cognitive tasks (e.g. logical reasoning tasks (Folkard 1975)) peaks in the late morning, and performance in short-term memory retention peaks in the early to mid-morning e.g. (Laird and McClumpha 1925) .These findings led to the hypothesis that the time of day at which a cognitive test is optimally completed is largely dependent on the specific parameters of the task, including the cognitive domain it belongs to, is duration and difficulty, the administration method, and the measured variable (Bonnet 2000). This and many other studies disclosed a temporal relationship between circadian variations in cognitive performance measures and daily fluctuations in physiological variables, such as when core body temperature (CBT) is high, neurobehavioral performance levels also tend to be high, whereas low CBT or high endogenous melatonin secretion are associated with reduced levels of neurobehavioral performance and alertness. (Kleitman, Titelbaum et al. 1938) explained this association by arguing that accuracy and speed in performance are contingent upon levels of muscle tonicity and in turn on the metabolic activity of the cells of the cerebral cortex. He therefore surmised that raising the latter through the circadian related increase in body temperature would indirectly speed up cognitive processing. This findings could help for another interpretation of our data such as that we found under SD lower levels of melatonin (Birchler-Pedross, Frey et al. 2011) and a tendency to higher core body temperature (Frey unpublished data) which could explain the faster reaction time. Although the circadian driven temperature oscillator is certainly one determinating factor of subjective vigilance, it is also influenced by the homeostatic process controlling the sleep-wake cycle.

CONCLUSION

Depressed women had faster RT in PVT performance during sleep deprivation. This may imply that depressed women are less susceptible to the wake-dependent detrimental aspects on PVT performance than non-depressed women. In other words, PVT performance in depression may not be directly linked to well-known "prefrontal tiredness", but rather involve subcortical structures and/or reflect higher cognitive arousal caused by rumination

ACKNOWLEDGEMENTS
We are very grateful to our technicians Claudia Renz, Marie-France Dattler, Giovanni Balestrieri and the student shift workers for their help in data acquisition. We thank to all the participants, especially the subject they came in our lab even suffering on a major depressive disorder. The authors thank Sarah Chellappa for Editing in English and for the fruitful discussions and comments on the manuscript and Dr. Antoine Viola for the suggestions and support in the statistical analysis.
This research was supported by Swiss National Science Foundation Grants START # 3100-055385.98, and 3130-0544991.98 and 320000-108108 to CC the Velux Foundation (Switzerland) and Bühlmann Laboratories, Allschwil (Switzerland).

CHAPTER 6

General Discussion

In this thesis, the theoretical framework that underpins depression is addressed within the context of the circadian and sleep-wake homeostatic regulation, and on how these two independent and non-additive systems impact mood, subjective sleepiness, melatonin and waking EEG in young depressive women.

Circadian and sleep-wake homeostatic repercussions on subjective well-being are age and gender dependent

Chapter 2 focuses on a model of how the circadian and homeostatic systems interact on subjective well-being, subjective sleepiness, melatonin and cortisol in young and older healthy subjects. I could successfully prove that our study design (40-h SD and 40-h nap protocol under constant routine conditions) was sensitive enough to quantify circadian and homeostatic influences on subjective well-being (i.e. composite score of mood, tension and physical comfort). Furthermore, I could show that the well known circadian regulation of subjective sleepiness, melatonin and cortisol are clearly age and gender dependent in healthy adult volunteers. Time of day modulation of subjective well-being was prominent in both the SD and the nap protocol participants, indicating that circadian phase plays a pivotal role in well-being. This is in accordance with data from a forced desynchrony protocol in which a significant component of mood regulation could be educed (Boivin, Czeisler et al. 1997). In general, and as a new interesting finding, both older adults and women were more affected by SD, showing a tendency to lower subjective well-being and a prominent circadian trough. The time course of subjective well-being displayed a significant circadian modulation, particularly in women under high sleep pressure conditions. Based on the results of young women and older volunteers, I hypothesized that the circadian dysregulation and sleep disturbances associated with depression may have profound detrimental effects on mood in depressed patients and older people, thus further perpetuating the disorder. However, this assumption could not be confirmed by results described in chapters 3 to 5. In chapter 2, subjective-well being was also compared with other circadian parameters such as sleepiness, cortisol and melatonin. These results indicate that subjective sleepiness and cortisol levels were significantly correlated with subjective well-being. Night time melatonin secretion in older volunteers was significantly lower than in the young, and all volunteers had slightly but significantly higher melatonin levels during the low (nap) than the high (SD) sleep pressure protocol. Although significant, the latter has probably no clinical relevance since the differences between SD and nap were rather small (< 2 pg/ml). However, in terms of melatonin, the most surprising result was that young women had significantly higher levels of nocturnal melatonin than young men and older men and women. The latter was not dependent on menstrual cycle, since another study in our laboratory yielded similar nocturnal melatonin levels in women studied during another phase of the menstrual cycle.

After the successful completion of the SD and the nap protocol in healthy young and older volunteers, we decided to apply exactly the same study design in young women suffering from an episode of major depression to test circadian and homeostatic related hypotheses that we formulated based on the previous literature (please see introduction) on depression.

I could identify many new and unexpected results regarding the influence of circadian and homeostatic effects on major depression (chapter 3 and 4). Some of the results of this thesis did not confirm the postulated hypothesis (please see Aims and Hypothesis of the Studies B). In the following I will briefly elucidate on the most interesting and surprising results and discuss them with respect to the previously postulated hypothesis and what it could imply for future treatments.

Possible impairments of sleep-wake homeostasis in depression measured by the EEG during wakefulness

In chapter 3, I focussed on the hypothesis that a deficiency in the homeostatic build-up of sleep pressure during sustained wakefulness (i.e. SD) could occur in depression, which in turn is responsible for less sleep intensity and thus less restful sleep. In contrast to a deficiency of the homeostatic process (i.e. S-deficiency), the results of this thesis rather indicated that depressed women might live on a higher level of homeostatic sleep pressure, as indexed by enhanced high frontal-low EEG activity (delta range from 2 to 5 Hz) during sustained wakefulness in the SD protocol. This finding of a higher sleep pressure in depressed was also supported by a tendency to have higher sleep pressure levels during the nap protocol ((Birchler-Pedross, unpublished data) illustrated in Figure Appendix 1). Along these line, we have evidence for significantly higher EEG slow-wave activity (delta EEG activity) levels in NREM sleep in the depressed cohort during baseline and recovery sleep in both the SD and the nap protocol (Frey, unpublished data).

Since mostly frontal brain areas were affected and these brain areas are particularly susceptible to the effects of sleep loss and quality of wakefulness, we interpret our results as reflecting a greater need for sleep, a different quality of wakefulness and/or less efficient sleep in our depressed women. For the latter however, we do not have statistical evidence, since sleep efficiency >90% was the same in both the control and the depressed women. Analysis of the microstructure of the sleep EEG, however, revealed that depressed women had more short-term arousals, which were not considered in the traditional sleep scoring analyses (analyses in progress). Thus, elevated slow-wave activity during NREM sleep could also reflect an immediate compensatory homeostatic response to these short-term arousals, a phenomenon, which was recently reported for primary insomnia (Parrino, Ferrillo et al. 2004).

Another underlying reason may be a use-dependent phenomenon, for example, the enhanced rumination in depression, which reinforces frontal delta activity. Although, in this project, we did not measure the degree of rumination via questionnaires, the depressed women expressed in general more concerns about the study protocol and every day matters than our control women.

In addition to the EEG, we also have evidence that the depressed women of this study were sleepier and tenser than the control women, thus contributing towards a significant effect during the biological night. Another unexpected finding was a significantly reduced night time secretion of salivary melatonin during sustained wakefulness in the depressed compared to the control women. This tendency to lower amplitude of melatonin could manifest a weaker circadian alerting signal, which manifests itself particularly in the late evening and thus allows more frontal delta activity. Another explanation could be that depressed women spend more time under artificial light conditions and less time under outdoor conditions than the controls prior to the study. Timing of melatonin onset (i.e. circadian phase) was not different compared to the control women. A possible explanation of the "non shift" in circadian phase might be an intact biological clock and/or the proposed selection of intermediate chronotypes for this study. Interestingly, we had to exclude many potential depressed study participants, due to a too late chronotype. Figure 6.1 illustrates the distribution of the interested volunteers. Thus, the study was biased in this regard, since the likelihood of being a late chronotype seems to be enhanced in depressed women. Interestingly, the randomly selected depressed subjects in the high sleep pressure protocol show in their distribution of intermediate chronotype a tendency to late type. This fact could have a profound influence on the amplitude of melatonin, which is lower during SD. Several studies, which found a phase-shift in depression, did not account for chronotypes. Thus, if a study determines phase shifts in depressed women or not, will strongly depend on the selection criteria.

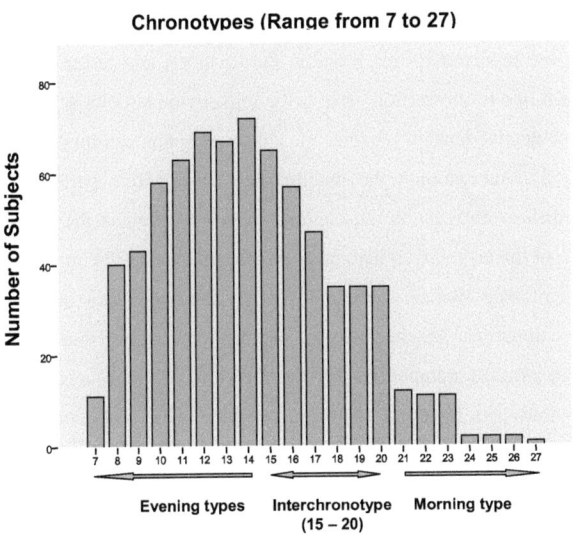

Figure 6.1: Illustration of the distribution of the chronotypes from the study applicants (unpublished data).

Diurnal mood variation and antidepressant effect of sleep deprivation

In chapter 4, I investigated if SD ameliorates mood as has been postulated in many studies (Wirz-Justice and Van den Hoofdakker 1999). The data show that depressed women without sleep disturbances show a significantly different time course of mood during the SD protocol (i.e. high sleep pressure) but not during nap protocol (i.e. low sleep pressure). Under SD they exhibited a more distinct circadian modulation of lower mood than controls. In their diurnal modulation they showed morning worsening and evening improvement, which corresponds to the so-called melancholic type. Despite this higher variability of mood fluctuation and characteristic of a more melancholic type as a predictor of SD response, surprisingly, and in contrast to my hypothesis, they did not profit from an antidepressant effect. These findings indicate that the possible profit of a treatment with SD could depend on the magnitude of insomnia in depression.

Neurobehavioral performance during sustained wakefulness in depression

In chapter 5 I focused on the influence of high and low sleep pressure conditions on neurobehavioral performance in depressed using the PVT. These results yielded an unexpected finding: depressed women had faster RT in PVT performance during SD. This may imply that depressed women are less susceptible to the wake-dependent aspects in PVT performance. This implies that PVT performance in depression may not be directly linked to well-known "prefrontal tiredness", which includes higher cognitive functions (executive, attention, memory), but rather that subcortical structures are involved in this process. Furthermore, this faster RT may also reflect higher cognitive arousal due to rumination seen in the higher frontal delta activity (chapter 3) and intensified by higher subjective tension (chapter 3). This assumption can be confirmed with results by Horne (Horne 1988). Accordingly, the maintenance of cognitive performance after SD is accompanied by a significant increase in muscle tension. Another aspect that should be taken into account is the duration of the task. Given that the PVT performance lasted only 5 minutes, it can be speculated that the depressed women performed faster since they can usually maintain their alertness for short time durations. Another possible explanation could be that increased motivation or interest can at least partially compensate for the effect of SD (Steyvers and Gaillard 1993). Unfortunately, in this study, we have not examined motivation by questionnaire. I do not have a true explanation for the faster PVT performance in depressed young women under high sleep pressure.

Conclusion and Perspectives

Taken together, this thesis provides several surprising data that conflict with some long-standing mainstream hypotheses (chapters 3 – 5). Overall, the homeostatic process seems to strongly influence young women with major depression disorder, as indicated by enhanced high frontal low-EEG activity. Higher subjective sleepiness and lower concentration of salivary melatonin amplitude during biological night of the depressed indicated that they did not profit from the antidepressant effect yet performed faster in PVT during SD. The waking EEG data suggests that the sleep-wake homeostat of the depressed is more an overdrive than an S-deficiency, which probably corresponds to the clinically reported day time fatigue. Depressed women who live on a higher level of sleep pressure do not profit from the antidepressant effect of SD and perform faster in a reaction time task than controls. This leads to the assumption that the excluded factors in this protocol (chronotype, insomnia) may exert a strong influence on the results. Considering the suggestion that hypocretin neurons stabilise arousal/alertness during periods of wakefulness and increase arousal-related behaviours, then it would be reasonable to regard it as an endogenous stressor. The heuristic model of emotional and physiological hyperarousal hypothesis (Basta, Chrousos et al. 2007) must be assumed, since the depression cohort in this study suffered only a minor hyperarousal, which probably affects frontal delta activity, leads to a higher sleep pressure but is not strong enough to cause sleep disturbances. The postulated abnormalities (i.e. phase-shift, sleep disturbances) in the biological rhythms seem to undergo less dramatic differences in the young depressed women who were investigated in this study; hence, it can be assumed that the homeostatic process seems to be more influential than previously thought.

Figure Annex 1:

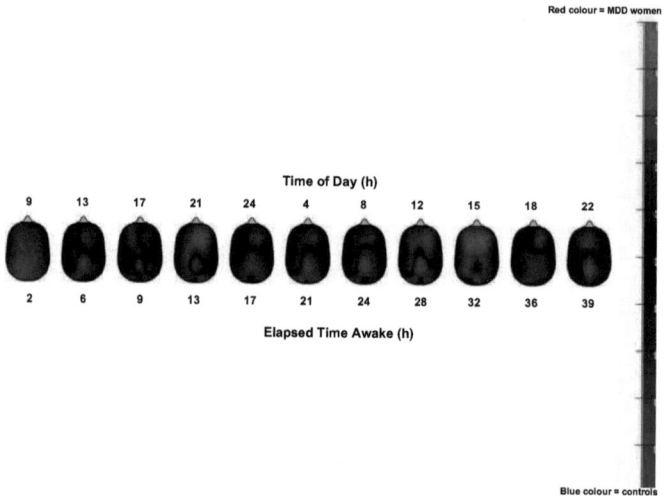

Figure Annex 1: Time course of low-frequency EEG activity (in the range of 0.5-5Hz), during the 40 h nap protocol (low sleep pressure). These contour plots (i.e., heads) describe the difference between low-frequency EEG activity in the MDD (red colour) and healthy control women (blue colour) over the 11 time points (Birchler-Pedross unpublished data).

REFERENCES

Achermann, P. and A. A. Borbély (1992). "Combining different models of sleep regulation." J Sleep Res 1(2): 144-147.

Achermann, P. and A. A. Borbély (2003). "Mathematical models of sleep regulation." Front Biosci 8: s683-693.

Achermann, P., D. J. Dijk, et al. (1993). "A model of human sleep homeostasis based on EEG slow-wave activity: quantitative comparison of data and simulations." Brain Res Bull 31(1-2): 97-113.

Adam, M., J. V. Retey, et al. (2006). "Age-related changes in the time course of vigilant attention during 40 hours without sleep in men." Sleep 29(1): 55-57.

Adan, A. and M. Sanchez-Turet (2001). "Gender differences in diurnal variations of subjective activation and mood." Chronobiol Int 18(3): 491-502.

Aeschbach, D., J. R. Matthews, et al. (1997). "Dynamics of the human EEG during prolonged wakefulness: evidence for frequency-specific circadian and homeostatic influences." Neurosci Lett 239(2-3): 121-124.

Aeschbach, D., J. R. Matthews, et al. (1999). "Two circadian rhythms in the human electroencephalogram during wakefulness." Am J Physiol 277(6 Pt 2): R1771-1779.

Aitken, R. C. (1969). "Measurement of feelings using visual analogue scales." Proc R Soc Med 62(10): 989-993.

Åkerstedt, T. and M. Gillberg (1990). "Subjective and objective sleepiness in the active individual." Int J Neurosci 52(1-2): 29-37.

Arendt, J. (2006). "Melatonin and human rhythms." Chronobiol Int 23(1-2): 21-37.

Armitage, R. (2007). "Sleep and circadian rhythms in mood disorders." Acta Psychiatr Scand Suppl: 104-115.

Armitage, R. (2007). "Sleep and circadian rhythms in mood disorders." Acta Psychiatr Scand Suppl(433): 104-115.

Armitage, R., R. Hoffmann, et al. (2000). "Slow-wave activity in NREM sleep: sex and age effects in depressed outpatients and healthy controls." Psychiatry Res 95(3): 201-213.

Aschoff, J. (1965). "Circadian Rhythms in Man." Science 148: 1427-1432.

Association, A. P. (1994). Diagnostic and Statistical Manual of Mental Disorders DSM-IV-TR

Aston-Jones, G., S. Chen, et al. (2001). "A neural circuit for circadian regulation of arousal." Nat Neurosci 4(7): 732-738.

Aston-Jones, G., J. Rajkowski, et al. (1999). "Role of locus coeruleus in attention and behavioral flexibility." Biol Psychiatry 46(9): 1309-1320.

Banks, S. and D. F. Dinges (2007). "Behavioral and physiological consequences of sleep restriction." J Clin Sleep Med 3(5): 519-528.

Basta, M., G. P. Chrousos, et al. (2007). "Chronic Insomnia and Stress System." Sleep Med Clin 2(2): 279-291.

Beaumont, M., D. Batejat, et al. (2001). "Slow release caffeine and prolonged (64-h) continuous wakefulness: effects on vigilance and cognitive performance." J Sleep Res 10(4): 265-276.

Beck, A. T., C. H. Ward, et al. (1961). "An inventory for measuring depression." Arch Gen Psychiatry 4: 561-571.

Beersma, D. G. (1998). "Models of human sleep regulation." Sleep Med Rev 2(1): 31-43.

Begleiter, H., B. Porjesz, et al. (1983). "P3 and stimulus incentive value." Psychophysiology 20(1): 95-101.

Belenky, G., N. J. Wesensten, et al. (2003). "Patterns of performance degradation and restoration during sleep restriction and subsequent recovery: a sleep dose-response study." J Sleep Res 12(1): 1-12.

Binks, P. G., W. F. Waters, et al. (1999). "Short-term total sleep deprivations does not selectively impair higher cortical functioning." Sleep 22(3): 328-334.

Birchler-Pedross, A. (unpublished data).

Birchler-Pedross, A., S. Frey, et al. (2011). "Higher frontal EEG synchronization in young women with major depression: a marker for increased homeostatic sleep pressure?" Sleep 34(12): 1699-1706.

Birchler-Pedross, A., S. Frey, et al. (2010). "Diurnal variations of mood in drug free unipolar depressed women under high and low sleep pressure conditions: is there a sleep deprivation effect ?" Sleep 33 (Abstract Supplement): A98.

Birchler-Pedross, A., C. M. Schröder, et al. (2009). "Subjective well-being is modulated by circadian phase, sleep pressure, age, and gender." J Biol Rhythms 24(3): 232-242.

Blatter, K., P. Graw, et al. (2006). "Gender and age differences in psychomotor vigilance performance under differential sleep pressure conditions." Behav Brain Res 168(2): 312-317.

Bliese, P. D., N. J. Wesensten, et al. (2006). "Age and individual variability in performance during sleep restriction." J Sleep Res 15(4): 376-385.

Boivin, D. B. (2000). "Influence of sleep-wake and circadian rhythm disturbances in psychiatric disorders." J Psychiatry Neurosci 25: 446-458.

Boivin, D. B., C. A. Czeisler, et al. (1997). "Complex interaction of the sleep-wake cycle and circadian phase modulates mood in healthy subjects." Arch Gen Psychiatry 54(2): 145-152.

Bonnet, II. M. (2000). Sleep deprivation. Philadelphia, W B Saunders.

Bonnet, M. H. (1989). "The effect of sleep fragmentation on sleep and performance in younger and older subjects." Neurobiol Aging 10(1): 21-25.

Bonnet, M. H. and D. L. Arand (2005). "Sleep latency testing as a time course measure of state arousal." J Sleep Res 14(4): 387-392.

Borbély, A. A. (1982). "A two process model of sleep regulation." Hum Neurobiol 1(3): 195-204.

Borbély, A. A. and P. Achermann (1992). "Concepts and models of sleep regulation: an overview." J Sleep Res 1(2): 63-79.

Borbély, A. A. and P. Achermann (1999). "Sleep homeostasis and models of sleep regulation." J Biol Rhythms 14(6): 557-568.

Borbély, A. A., F. Baumann, et al. (1981). "Sleep deprivation: effect on sleep stages and EEG power density in man." Electroencephalogr Clin Neurophysiol 51(5): 483-495.

Borbély, A. A., P. Steigrad, et al. (1980). "Effect of sleep deprivation on brain serotonin in the rat." Behav Brain Res 1(2): 205-210.

Borbély, A. A. and I. Tobler (1989). "Endogenous sleep-promoting substances and sleep regulation." Physiol Rev 69(2): 605-670.

Borbély, A. A. and A. Wirz-Justice (1982). "Sleep, sleep deprivation and depression. A hypothesis derived from a model of sleep regulation." Hum Neurobiol 1(3): 205-210.

Brendel, D. H., C. F. Reynolds, 3rd, et al. (1990). "Sleep stage physiology, mood, and vigilance responses to total sleep deprivation in healthy 80-year-olds and 20-year-olds." Psychophysiology 27(6): 677-685.

Breslau, N., T. Roth, et al. (1996). "Sleep disturbance and psychiatric disorders: a longitudinal epidemiological study of young adults." Biol Psychiatry 39(6): 411-418.

Brun, J., G. Chamba, et al. (1998). "Effect of modafinil on plasma melatonin, cortisol and growth hormone rhythms, rectal temperature and performance in healthy subjects during a 36 h sleep deprivation." J Sleep Res 7(2): 105-114.

Brunner, D. P., D. J. Dijk, et al. (1993). "Repeated partial sleep deprivation progressively changes in EEG during sleep and wakefulness." Sleep 16(2): 100-113.

Buijs, R. M., F. A. Scheer, et al. (2006). "Organization of circadian functions: interaction with the body." Prog Brain Res 153: 341-360.

Bunney, W. E., Jr., D. L. Murphy, et al. (1970). "The switch process from depression to mania: relationship to drugs which alter brain amines." Lancet 1(7655): 1022-1027.

Buysse, D. J., E. Frank, et al. (1997). "Electroencephalographic sleep correlates of episode and vulnerability to recurrence in depression." Biol Psychiatry 41(4): 406-418.

Buysse, D. J., D. J. Kupfer, et al. (1999). "Effects of prior fluoxetine treatment on EEG sleep in women with recurrent depression." Neuropsychopharmacology 21(2): 258-267.

Buysse, D. J., T. H. Monk, et al. (1995). "Circadian patterns of unintended sleep episodes during a constant routine in remitted depressed patients." J Psychiatr Res 29: 407-416.

Buysse, D. J., T. H. Monk, et al. (1993). "Patterns of sleep episodes in young and elderly adults during a 36-hour constant routine." Sleep 16(7): 632-637.

Buysse, D. J., C. F. Reynolds, 3rd, et al. (1989). "The Pittsburgh Sleep Quality Index: a new instrument for psychiatric practice and research." Psychiatry Res 28(2): 193-213.

Cajochen, C., R. Foy, et al. (1999). "Frontal predominance of a relative increase in sleep delta and theta EEG activity after sleep loss in humans." Sleep Res Online 2(3): 65-69.

Cajochen, C., S. B. Khalsa, et al. (1999). "EEG and ocular correlates of circadian melatonin phase and human performance decrements during sleep loss." Am J Physiol 277(3 Pt 2): R640-649.

Cajochen, C., V. Knoblauch, et al. (2001). "Dynamics of frontal EEG activity, sleepiness and body temperature under high and low sleep pressure." Neuroreport 12(10): 2277-2281.

Cajochen, C., V. Knoblauch, et al. (2004). "Circadian modulation of sequence learning under high and low sleep pressure conditions." Behav Brain Res 151(1-2): 167-176.

Cajochen, C., M. Münch, et al. (2006). "Age-related changes in the circadian and homeostatic regulation of human sleep." Chronobiol Int 23(1-2): 461-474.

Cajochen, C., J. K. Wyatt, et al. (2002). "Separation of circadian and wake duration-dependent modulation of EEG activation during wakefulness." Neuroscience 114(4): 1047-1060.

Caldwell, J. A. and S. R. Gilreath (2002). "A survey of aircrew fatigue in a sample of U.S. Army aviation personnel." Aviat Space Environ Med 73(5): 472-480.

Caldwell, J. A., Jr. and P. A. LeDuc (1998). "Gender influences on performance, mood and recovery sleep in fatigued aviators." Ergonomics 41(12): 1757-1770.

Carskadon, M. A. and W. C. Dement (1982). "Nocturnal determinants of daytime sleepiness." Sleep 5 Suppl 2: S73-81.

Casper, R. C., D. E. Redmond, Jr., et al. (1985). "Somatic symptoms in primary affective disorder. Presence and relationship to the classification of depression." Arch Gen Psychiatry 42(11): 1098-1104.

Cassone, V. M. and A. K. Natesan (1997). "Time and time again: the phylogeny of melatonin as a transducer of biological time." J Biol Rhythms 12(6): 489-497.

Cheng, P., J. Goldschmied, et al. (2010). "Slow wave sleep enhancement and positie mood in depressive and controls." Sleep 33 (Abstract Supplement): A234.

Cole, M. G. and N. Dendukuri (2003). "Risk factors for depression among elderly community subjects: a systematic review and meta-analysis." Am J Psychiatry 160(6): 1147-1156.

Colquhoun, P. (1981). Rhythms in performance. Handbook of behavioral neurobiology. J. Aschoff. New York, Plenum Press. 4.

Covi, L., R. Lipman, et al. (1977). "Transcultural comparability of outpatient self-ratings: II. The profile of mood states (POMS)." Drugs and Experimental Clinical Research 1: 141-151.

Curran-Everett, D. (2000). "Multiple comparisons: philosophies and illustrations." Am J Physiol Regul Integr Comp Physiol 279(1): R1-8.

Czeisler, C. A., J. S. Allan, et al. (1986). "Bright light resets the human circadian pacemaker independent of the timing of the sleep-wake cycle." Science 233(4764): 667-671.

Daan, S., D. G. Beersma, et al. (1984). "Timing of human sleep: recovery process gated by a circadian pacemaker." Am J Physiol 246(2 Pt 2): R161-183.

Danilenko, K. V., C. Cajochen, et al. (2003). "Is sleep per se a zeitgeber in humans?" J Biol Rhythms 18(2): 170-178.

Desan, P. H., D. A. Oren, et al. (2000). "Genetic polymorphism at the CLOCK gene locus and major depression." Am J Med Genet 96(3): 418-421.

Dijk, D. J., D. G. Beersma, et al. (1987). "EEG power density during nap sleep: reflection of an hourglass measuring the duration of prior wakefulness." J Biol Rhythms 2(3): 207-219.

Dijk, D. J. and C. A. Czeisler (1995). "Contribution of the circadian pacemaker and the sleep homeostat to sleep propensity, sleep structure, electroencephalographic slow waves, and sleep spindle activity in humans." J Neurosci 15(5 Pt 1): 3526-3538.

Dinges, D. F., F. Pack, et al. (1997). "Cumulative sleepiness, mood disturbance, and psychomotor vigilance performance decrements during a week of sleep restricted to 4-5 hours per night." Sleep 20(4): 267-277.

Doran, S. M., H. P. Van Dongen, et al. (2001). "Sustained attention performance during sleep deprivation: evidence of state instability." Arch Ital Biol 139(3): 253-267.

Drevets, W. C. (2001). "Neuroimaging and neuropathological studies of depression: implications for the cognitive-emotional features of mood disorders." Curr Opin Neurobiol 11(2): 240-249.

Drummond, S. P., A. Bischoff-Grethe, et al. (2005). "The neural basis of the psychomotor vigilance task." Sleep 28(9): 1059-1068.

Drummond, S. P., M. J. Meloy, et al. (2005). "Compensatory recruitment after sleep deprivation and the relationship with performance." Psychiatry Res 140(3): 211-223.

Dryman, A. and W. W. Eaton (1991). "Affective symptoms associated with the onset of major depression in the community: findings from the US National Institute of Mental Health Epidemiologic Catchment Area Program." Acta Psychiatr Scand 84(1): 1-5.

Duffy, J. F. and D. J. Dijk (2002). "Getting through to circadian oscillators: why use constant routines?" J Biol Rhythms 17(1): 4-13.

Ebisawa, T. (2007). "Circadian rhythms in the CNS and peripheral clock disorders: human sleep disorders and clock genes." J Pharmacol Sci 103(2): 150-154.

Edgar, D. M., W. C. Dement, et al. (1993). "Effect of SCN lesions on sleep in squirrel monkeys: evidence for opponent processes in sleep-wake regulation." J Neurosci 13(3): 1065-1079.

Finelli, L. A., H. Baumann, et al. (2000). "Dual electroencephalogram markers of human sleep homeostasis: correlation between theta activity in waking and slow-wave activity in sleep." Neuroscience 101(3): 523-529.

Finelli, L. A., A. A. Borbély, et al. (2001). "Functional topography of the human nonREM sleep electroencephalogram." Eur J Neurosci 13(12): 2282-2290.

Folkard, S. (1975). "Diurnal variation in logical reasoning." Br J Psychol 66(1): 1-8.

Folstein, M. F. and R. Luria (1973). "Reliability, validity, and clinical application of the Visual Analogue Mood Scale." Psychol Med 3(4): 479-486.

Franken, P., D. Chollet, et al. (2001). "The homeostatic regulation of sleep need is under genetic control." J Neurosci 21(8): 2610-2621.

Frey, D. J., P. Badia, et al. (2004). "Inter- and intra-individual variability in performance near the circadian nadir during sleep deprivation." J Sleep Res 13(4): 305-315.

Frey, S. (unpublished data).

Frey, S., A. Birchler-Pedross, et al. (2010). "Women with major depression live under higher homeostatic sleep pressure." Journal of Sleep Research: Abstract P441.

Fuchs, E., M. Simon, et al. (2006). "Pharmacology of a new antidepressant: benefit of the implication of the melatonergic system." Int Clin Psychopharmacol 21 Suppl 1: S17-20.

Germain, A. and D. J. Kupfer (2008). "Circadian rhythm disturbances in depression." Hum Psychopharmacol 23: 571-585.

Giedke, H., H. Wormstall, et al. (1990). "Therapeutic sleep deprivation in depressives, restricted to the two nocturnal hours between 3:00 and 5:00." Prog Neuropsychopharmacol Biol Psychiatry 14(1): 37-47.

Gillberg, M., G. Kecklund, et al. (1994). "Relations between performance and subjective ratings of sleepiness during a night awake." Sleep 17(3): 236-241.

Gillin, J. C., W. Duncan, et al. (1979). "Successful separation of depressed, normal, and insomniac subjects by EEG sleep data." Arch Gen Psychiatry 36(1): 85-90.

Graw, P., H. J. Haug, et al. (1998). "Sleep deprivation response in seasonal affective disorder during a 40-h constant routine." J Affect Disord 48(1): 69-74.

Graw, P., K. Kräuchi, et al. (2004). "Circadian and wake-dependent modulation of fastest and slowest reaction times during the psychomotor vigilance task." Physiol Behav 80(5): 695-701.

Harris, C. D. (2005). "Neurophysiology of sleep and wakefulness." Respir Care Clin N Am 11(4): 567-586.

Harrison, Y. and J. A. Horne (1998). "Sleep loss impairs short and novel language tasks having a prefrontal focus." J Sleep Res 7(2): 95-100.

Harrison, Y., J. A. Horne, et al. (2000). "Prefrontal neuropsychological effects of sleep deprivation in young adults--a model for healthy aging?" Sleep 23(8): 1067-1073.

Haug, H. J. (1992). "Prediction of sleep deprivation outcome by diurnal variation of mood." Biol Psychiatry 31(3): 271-278.

Haug, H. J. and A. Wirz-Justice (1993). "Diurnal variation of mood in depression: important or irrelevant?" Biol Psychiatry 34(4): 201-203.

Hendrickson, A. E., N. Wagoner, et al. (1972). "An autoradiographic and electron microscopic study of retino-hypothalamic connections." Z Zellforsch Mikrosk Anat 135(1): 1-26.

Homberg, V., G. Grunewald, et al. (1981). "The variation of p300 amplitude in a money-winning paradigm in children." Psychophysiology 18(3): 258-262.

Horne, J. A. (1988). "Sleep loss and "divergent" thinking ability." Sleep 11(6): 528-536.

Horne, J. A. (1993). "Human sleep, sleep loss and behaviour. Implications for the prefrontal cortex and psychiatric disorder." Br J Psychiatry 162: 413-419.

Horne, J. A. and L. A. Reyner (1995). "Sleep related vehicle accidents." Bmj 310(6979): 565-567.

Hull, J. T., K. P. Wright, Jr., et al. (2003). "The influence of subjective alertness and motivation on human performance independent of circadian and homeostatic regulation." J Biol Rhythms 18(4): 329-338.

Janet, B. W., D. S. W. Williams, et al. (2003). Structured Interview Guide For The Hamilton Depression Rating Scale with atypical depression supplement (SIGH-ADS).

Kalsbeek, A., I. F. Palm, et al. (2006). "SCN outputs and the hypothalamic balance of life." J Biol Rhythms 21(6): 458-469.

Kleitman, N., S. Titelbaum, et al. (1938). "The effect of body temperature on reaction time." American ournal of Physiology(121): 495-501.

Knoblauch, V., W. Martens, et al. (2003). "Regional differences in the circadian modulation of human sleep spindle characteristics." Eur J Neurosci 18(1): 155-163.

Knoblauch, V., W. L. Martens, et al. (2003). "Human sleep spindle characteristics after sleep deprivation." Clin Neurophysiol 114(12): 2258-2267.

Knoblauch, V., M. Münch, et al. (2005). "Age-related changes in the circadian modulation of sleep-spindle frequency during nap sleep." Sleep 28(9): 1093-1101.

Knott, V. J. and Y. D. Lapierre (1987). "Computerized EEG correlates of depression and antidepressant treatment." Prog Neuropsychopharmacol Biol Psychiatry 11(2-3): 213-221.

Koorengevel, K. M., D. G. Beersma, et al. (2002). "A forced desynchrony study of circadian pacemaker characteristics in seasonal affective disorder." J Biol Rhythms 17(5): 463-475.

Koorengevel, K. M., D. G. Beersma, et al. (2003). "Mood regulation in seasonal affective disorder patients and healthy controls studied in forced desynchrony." Psychiatry Res 117(1): 57-74.

Kupfer, D. J. and F. G. Foster (1972). "Interval between onset of sleep and rapid-eye-movement sleep as an indicator of depression." Lancet 2(7779): 684-686.

Kupfer, D. J., R. F. Ulrich, et al. (1984). "Application of automated REM and slow wave sleep analysis: II. Testing the assumptions of the two-process model of sleep regulation in normal and depressed subjects." Psychiatry Res 13(4): 335-343.

Laird, D. A. and T. McClumpha (1925). "Sex Difference, in Emotional Outlets." Science 62(1604): 292.

Landolt, H, P,, D. J. Dijk, et al. (1995). "Caffeine reduces low-frequency delta activity in the human sleep EEG." Neuropsychopharmacology 12(3): 229-238.

Landolt, H. P. and J. C. Gillin (2005). "Similar sleep EEG topography in middle-aged depressed patients and healthy controls." Sleep 28(2): 239-247.

Landsness, E. C., M. R. Goldstein, et al. (2010). "Selective slow wave deprivation as a possible acute treament of major depressive disorder." Sleep 33 (Abstract Supplement): A234.

Latini, S. and F. Pedata (2001). "Adenosine in the central nervous system: release mechanisms and extracellular concentrations." J Neurochem 79(3): 463-484.

Lavie, P. (1986). "Ultrashort sleep-waking schedule. III. 'Gates' and 'forbidden zones' for sleep." Electroencephalogr Clin Neurophysiol 63(5): 414-425.

Lavie, P. (2001). "Sleep-wake as a biological rhythm." Annu Rev Psychol 52: 277-303.

Leproult, R., E. F. Colecchia, et al. (2003). "Individual differences in subjective and objective alertness during sleep deprivation are stable and unrelated." Am J Physiol Regul Integr Comp Physiol 284(2): R280-290.

Leproult, R., O. Van Reeth, et al. (1997). "Sleepiness, performance, and neuroendocrine function during sleep deprivation: effects of exposure to bright light or exercise." J Biol Rhythms 12(3): 245-258.

Lustberg, L. and C. F. Reynolds (2000). "Depression and insomnia: questions of cause and effect." Sleep Med Rev 4(3): 253-262.

Makeig, S., T. P. Jung, et al. (2000). "Awareness during drowsiness: dynamics and electrophysiological correlates." Can J Exp Psychol 54(4): 266-273.

Mallon, L., J. E. Broman, et al. (2000). "Relationship between insomnia, depression, and mortality: a 12-year follow-up of older adults in the community." Int Psychogeriatr 12(3): 295-306.

Mayberg, H. S. (2003). "Positron emission tomography imaging in depression: a neural systems perspective." Neuroimaging Clin N Am 13(4): 805-815.

Maywood, E. S., J. O'Neill, et al. (2006). "Circadian timing in health and disease." Prog Brain Res 153: 253-269.

McCarley, R. W. (2007). "Neurobiology of REM and NREM sleep." Sleep Med 8(4): 302-330.

McClung, C. A. (2007). "Circadian genes, rhythms and the biology of mood disorders." Pharmacol Ther 114(2): 222-232.

McIntyre, R. S., J. Z. Konarski, et al. (2005). "Measuring the severity of depression and remission in primary care: validation of the HAMD-7 scale." CMAJ 173(11): 1327-1334.

McNair, D., M. Lorr, et al. (1971). "Manual for the Profile of Mood states." San Diego: Educational and Industrial Testing Service.

Mignot, E., S. Taheri, et al. (2002). "Sleeping with the hypothalamus: emerging therapeutic targets for sleep disorders." Nat Neurosci 5 Suppl: 1071-1075.

Mills, J. N., D. S. Minors, et al. (1978). "Adaptation to abrupt time shifts of the oscillator(s) controlling human circadian rhythms." J Physiol 285: 455-470.

Moller, H. J., G. M. Devins, et al. (2006). "Sleepiness is not the inverse of alertness: evidence from four sleep disorder patient groups." Exp Brain Res 173(2): 258-266.

Monk, T. H. (1982). "The arousal model of time of day effects in human performance efficiency." Chronobiologia 9(1): 49-54.

Monk, T. H. (1989). "A Visual Analogue Scale technique to measure global vigor and affect." Psychiatry Res 27(1): 89-99.

Monk, T. H., D. J. Buysse, et al. (1992). "Rhythmic vs homeostatic influences on mood, activation, and performance in young and old men." J Gerontol 47(4): P221-227.

Montgomery, S. A. and M. Åsberg (1979). "A new depression scale designed to be sensitive to change." Br J Psychiatry 134: 382-389.

Moore, R. Y. (1973). "Retinohypothalamic projection in mammals: a comparative study." Brain Res 49(2): 403-409.

Moore, R. Y. (2007). "Suprachiasmatic nucleus in sleep-wake regulation." Sleep Med 8 Suppl 3: 27-33.

Moore, R. Y. and N. J. Lenn (1972). "A retinohypothalamic projection in the rat." J Comp Neurol 146(1): 1-14.

Münch, M., V. Knoblauch, et al. (2004). "The frontal predominance in human EEG delta activity after sleep loss decreases with age." Eur J Neurosci 20(5): 1402-1410.

Münch, M., V. Knoblauch, et al. (2005). "Age-related attenuation of the evening circadian arousal signal in humans." Neurobiol Aging 26(9): 1307-1319.

Münch, M., V. Knoblauch, et al. (2007). "Is homeostatic sleep regulation under low sleep pressure modified by age?" Sleep 30(6): 781-792.

Murray, G., N. B. Allen, et al. (2002). "Mood and the circadian system: investigation of a circadian component in positive affect." Chronobiol Int 19(6): 1151-1169.

Nestler, E. J., M. Barrot, et al. (2002). "Neurobiology of depression." Neuron 34(1): 13-25.

Nissen, C., B. Feige, et al. (2001). "Delta sleep ratio as a predictor of sleep deprivation response in major depression." J Psychiatr Res 35(3): 155-163.

Nissen, C., J. Holz, et al. (2010). "Learning as a model for neural plasticity in major depression." Biol Psychiatry 68(6): 544-552.

Nolen-Hoeksema, S. (1991). "Responses to depression and their effects on the duration of depressive episodes." J Abnorm Psychol 100(4): 569-582.

Nolen-Hoeksema, S. (2000). "The role of rumination in depressive disorders and mixed anxiety/depressive symptoms." J Abnorm Psychol 109(3): 504-511.

Ohayon, M. M. and T. Roth (2003). "Place of chronic insomnia in the course of depressive and anxiety disorders." J Psychiatr Res 37(1): 9-15.

Pace-Schott, E. F. and J. A. Hobson (2002). "The neurobiology of sleep: genetics, cellular physiology and subcortical networks." Nat Rev Neurosci 3(8): 591-605.

Pack, A. I., A. M. Pack, et al. (1995). "Characteristics of crashes attributed to the driver having fallen asleep." Accid Anal Prev 27(6): 769-775.

Papousek, M. (1975). "[Chronobiological aspects of cyclothymia (author's transl)]." Fortschr Neurol Psychiatr Grenzgeb 43(8): 381-440.

Parrino, L., F. Ferrillo, et al. (2004). "Is insomnia a neurophysiological disorder? The role of sleep EEG microstructure." Brain Res Bull 63(5): 377-383.

Perlis, M. L., D. E. Giles, et al. (1997). "Psychophysiological insomnia: the behavioural model and a neurocognitive perspective." J Sleep Res 6(3): 179-188.

Philip, P., P. Sagaspe, et al. (2005). "Fatigue, sleepiness, and performance in simulated versus real driving conditions." Sleep 28(12): 1511-1516.

Philip, P., J. Taillard, et al. (2004). "Age, performance and sleep deprivation." J Sleep Res 13(2): 105-110.

Pollock, V. and L. Schneider (1991). "Quantitative waking EEG in depression. ." Biol Psychiatry 27: 757-780.

Porkka-Heiskanen, T., R. E. Strecker, et al. (1997). "Adenosine: a mediator of the sleep-inducing effects of prolonged wakefulness." Science 276(5316): 1265-1268.

Ramos-Brieva, J. and A. Cordero-Villafafila (1988). "A new validation of the Hamilton Rating Scale for Depression." J Psychiatr Res 22: 21-28.

Reinink, E., N. Bouhuys, et al. (1990). "Prediction of the antidepressant response to total sleep deprivation by diurnal variation of mood." Psychiatry Res 32(2): 113-124.

Retey, J. V., M. Adam, et al. (2005). "A functional genetic variation of adenosine deaminase affects the duration and intensity of deep sleep in humans." Proc Natl Acad Sci U S A 102(43): 15676-15681.

Reynolds, C. F., 3rd, D. J. Kupfer, et al. (1986). "Sleep deprivation in healthy elderly men and women: effects on mood and on sleep during recovery." Sleep 9(4): 492-501.

Riemann, D., M. Berger, et al. (2001). "Sleep and depression--results from psychobiological studies: an overview." Biol Psychol 57(1-3): 67-103.

Riemann, D. and U. Voderholzer (2003). "Primary insomnia: a risk factor to develop depression?" J Affect Disord 76(1-3): 255-259.

Robbins, T. W. and B. J. Everitt (1999). "Drug addiction: bad habits add up." Nature 398(6728): 567-570.

Rogers, N. L., J. Dorrian, et al. (2003). "Sleep, waking and neurobehavioural performance." Front Biosci 8: s1056-1067.

Rosenzweig-Lipson, S., C. E. Beyer, et al. (2007). "Differentiating antidepressants of the future: efficacy and safety." Pharmacol Ther 113(1): 134-153.

Roybal, K., D. Theobold, et al. (2007). "Mania-like behavior induced by disruption of CLOCK." Proc Natl Acad Sci U S A 104(15): 6406-6411.

Saper, C. B., T. C. Chou, et al. (2001). "The sleep switch: hypothalamic control of sleep and wakefulness." Trends Neurosci 24(12): 726-731.

Scheen, A. J., M. M. Byrne, et al. (1996). "Relationships between sleep quality and glucose regulation in normal humans." Am J Physiol 271(2 Pt 1): E261-270.

Schmidt, C., F. Collette, et al. (2009). "Homeostatic sleep pressure and responses to sustained attention in the suprachiasmatic area." Science 324(5926): 516-519.

Scott, J. P., L. R. McNaughton, et al. (2006). "Effects of sleep deprivation and exercise on cognitive, motor performance and mood." Physiol Behav 87(2): 396-408.

Selvi, Y., M. Gulec, et al. (2007). "Mood changes after sleep deprivation in morningness-eveningness chronotypes in healthy individuals." J Sleep Res 16(3): 241-244.

Shirani, A. and E. K. St Louis (2009). "Illuminating rationale and uses for light therapy." J Clin Sleep Med 5(2): 155-163.

Souetre, E., E. Salvati, et al. (1989). "Circadian rhythms in depression and recovery: evidence for blunted amplitude as the main chronobiological abnormality." Psychiatry Res 28: 263-278.

Srinivasan, V., M. Smits, et al. (2006). "Melatonin in mood disorders." World J Biol Psychiatry 7(3): 138-151.

Steyvers, F. J. and A. W. Gaillard (1993). "The effects of sleep deprivation and incentives on human performance." Psychol Res 55(1): 64-70.

Strogatz, S. H., R. E. Kronauer, et al. (1987). "Circadian pacemaker interferes with sleep onset at specific times each day: role in insomnia." Am J Physiol 253(1 Pt 2): R172-178.

Sylvester, C. M., K. E. Krout, et al. (2002). "Suprachiasmatic nucleus projection to the medial prefrontal cortex: a viral transneuronal tracing study." Neuroscience 114(4): 1071-1080.

Taub, J. and R. Berger (1974). Effects of acute shifts in circadian rhythms of sleep and wakefulness on performance and mood, Scheving, LE, Halberg, F, Pauly, JE.

Taub, J. M. and R. J. Berger (1973). "Performance and mood following variations in the length and timing of sleep." Psychophysiology 10(6). 559-570.

Thomas, M., H. Sing, et al. (2000). "Neural basis of alertness and cognitive performance impairments during sleepiness. I. Effects of 24 h of sleep deprivation on waking human regional brain activity." J Sleep Res 9(4): 335-352.

Tobler, I. and A. A. Borbély (1986). "Sleep EEG in the rat as a function of prior waking." Electroencephalogr Clin Neurophysiol 64(1): 74-76.

Tononi, G. (2009). "Slow wave homeostasis and synaptic plasticity." J Clin Sleep Med 5(2 Suppl): S16-19.

Torsvall, L. and T. Åkerstedt (1980). "A diurnal type scale. Construction, consistency and validation in shift work." Scand J Work Environ Health 6(4): 283-290.

Torsvall, L. and T. Åkerstedt (1987). "Sleepiness on the job: continuously measured EEG changes in train drivers." Electroencephalogr Clin Neurophysiol 66(6): 502-511.

Totterdell, P., S. Reynolds, et al. (1994). "Associations of sleep with everyday mood, minor symptoms and social interaction experience." Sleep 17(5): 466-475.

Van Den Burg, W., D. G. Beersma, et al. (1992). "Self-rated arousal concurrent with the antidepressant response to total sleep deprivation of patients with a major depressive disorder: a disinhibition hypothesis." J Sleep Res 1(4): 211-222.

Van den Hoofdakker, R. H., D. G. Beersma, et al. (1986). "Sleep disorders in depression." Eur Neurol 25 Suppl 2: 66-70.

Van Dongen, H. P. and D. F. Dinges (2005). "Sleep, circadian rhythms, and psychomotor vigilance." Clin Sports Med 24(2): 237-249, vii-viii.

van Eekelen, A. P., G. A. Kerkhof, et al. (2003). "Circadian variation in cortisol reactivity to an acute stressor." Chronobiol Int 20(5): 863-878.

Von Zerssen, D. and D. M. Koeller (1976). Paranoid-Depressivitätsskala. Weinheim, Beltz Testgesellschaft.

Voultsios, A., D. J. Kennaway, et al. (1997). "Salivary melatonin as a circadian phase marker: validation and comparison to plasma melatonin." J Biol Rhythms 12(5): 457-466.

Watson, D. and L. A. Clark (1997). "Measurement and mismeasurement of mood: recurrent and emergent issues." J Pers Assess 68(2): 267-296.

Webb, W. B. and H. W. Agnew, Jr. (1975). "Sleep efficiency for sleep-wake cycles of varied length." Psychophysiology 12(6): 637-641.

Weber, J., S. JC, et al. (1997). "A direct ultrasensitive RIA for the determination of melatonin in human saliva: comparison with serum levels. ." J Sleep Res 26: 757.

Wehr, T. A. and H. K. H. Goodwin (1981). Biological rhythms and psychiatry. American Handbook of Psychiatry. B. H. Arieti S. New York, Basic Books. 17: 46-74.

Wehr, T. A. and A. Wirz-Justice (1982). "Circadian rhythm mechanisms in affective illness and in antidepressant drug action." Pharmacopsychiatria 15(1): 31-39.

Weingartner, H., R. M. Cohen, et al. (1981). "Cognitive processes in depression." Arch Gen Psychiatry 38(1): 42-47.

Werth, E., P. Achermann, et al. (1997). "Fronto-occipital EEG power gradients in human sleep." J Sleep Res 6(2): 102-112.

Wirz-Justice, A. (1995). Biological rhythms in mood disorders. Psychopharmacology: thr Fourth Generation of Progress. F. E. Bloom and D. J. Kupfer. New York, Raven Press: 999-1017.

Wirz-Justice, A. (1998). "Why is sleep deprivation an orphan drug?" Psychiatry Res 81(2): 281-282.

Wirz-Justice, A. (2003). "Chronobiology and mood disorders." Dialogues Clin Neurosci 5: 315-325.

Wirz-Justice, A. (2006). "Biological rhythm disturbances in mood disorders." Int Clin Psychopharmacol 21 Suppl 1: S11-15.

Wirz-Justice, A. (unpublished figure).

Wirz-Justice, A., F. Benedetti, et al. (2005). "Chronotherapeutics (light and wake therapy) in affective disorders." Psychol Med 35(7): 939-944.

Wirz-Justice, A., I. Tobler, et al. (1981). "Sleep deprivation: effects on circadian rhythms of rat brain neurotransmitter receptors." Psychiatry Res 5(1): 67-76.

Wirz-Justice, A. and R. H. Van den Hoofdakker (1999). "Sleep deprivation in depression: what do we know, where do we go?" Biol Psychiatry 46(4): 445-453.

Wittchen, H. U., U. Wunderlich, et al. (1997). SKID-I-Structured clinical interview for DSM-IV. Göttingen, Hogrefe.

Wittchen, H. U., U. Wunderlich, et al. (1996). Strukturiertes Klinisches Interview für DSM-IV (SKID). Göttingen, Beltz-Test.

Wood, C. and M. E. Magnello (1992). "Diurnal changes in perceptions of energy and mood." J R Soc Med 85(4): 191-194.

Wright, K. P., Jr., R. J. Hughes, et al. (2001). "Intrinsic near-24-h pacemaker period determines limits of circadian entrainment to a weak synchronizer in humans." Proc Natl Acad Sci U S A 98(24): 14027-14032.

Wyatt, J. K., A. Ritz-De Cecco, et al. (1999). "Circadian temperature and melatonin rhythms, sleep, and neurobehavioral function in humans living on a 20-h day." Am J Physiol 277(4 Pt 2): R1152-1163.

List of Figures

Figure 1.1: Schematic representation of the homeostatic and circadian process	8
Figure 1.2: Schematic representation of the circadian system	10
Figure 1.3: Schematic figure of the "opponent processes"	11
Figure 1.4: Circadian and wake-dependent variations of mood	12
Figure 1.5: Schematic representation of the two-process model of sleep regulation	16
Figure 1.6: Schematic representation of the two 3.5-day protocols	18
Figure 2.1: Overview of the 3.5-day laboratory protocol	28
Figure 2.2: Time course of subjective well-being (low sleep pressure)	33
Figure 2.3: Time course of subjective well-being (high sleep pressure)	35
Figure 2.4: Time course of the correlation coefficient (r^2)	37
Figure 3.1: Absolute EEG power spectra during extended wakefulness	49
Figure 3.2: Dynamics of low-frequency EEG activity during extended wakefulness	50
Figure 3.3: Colour-coded scalp topography of low-frequency EEG power density	50
Figure 3.4: Time course of low-frequency EEG activity during extended wakefulness	51
Figure 3.5: Time course of subjective sleepiness, subjective tension and salivary melatonin	52
Figure 3.6: Linear regression between the values of the depression sore ante the value of FLA	53
Figure 4.1: Time course of subjective mood	63
Figure 4.2: Estimation of depression levels by observer ratings using the Hamilton-7 Items	64
Figure 4.3: Linear regression between the values of PSQI and the values of mood	65
Figure 5.1: Median RT (ms) of the depressed and control group	74
Figure 5.2: Time course of vigilance performance	75
Figure 6.1: Illustration of the distribution of the chronotypes	83
Figure Annex 1: Time course of low-frequency EEG (low sleep pressure protocol)	86

List of Tables

Table 2.1: Main and interaction effects on subjective well-being and the other parameters	30/31
Table 2.2: Results of the backward stepwise regression analysis	36
Table 5.1: Results of the r ANOVAs of the PVT	74

i want morebooks!

Buy your books fast and straightforward online - at one of world's fastest growing online book stores! Environmentally sound due to Print-on-Demand technologies.

Buy your books online at
www.get-morebooks.com

Kaufen Sie Ihre Bücher schnell und unkompliziert online – auf einer der am schnellsten wachsenden Buchhandelsplattformen weltweit! Dank Print-On-Demand umwelt- und ressourcenschonend produziert.

Bücher schneller online kaufen
www.morebooks.de

 VDM Verlagsservicegesellschaft mbH
Heinrich-Böcking-Str. 6-8　　Telefon: +49 681 3720 174　　info@vdm-vsg.de
D - 66121 Saarbrücken　　　Telefax: +49 681 3720 1749　　www.vdm-vsg.de

Printed by Books on Demand GmbH, Norderstedt / Germany